Hot

THE INSIDE STORY OF THE BATTLE AGAINST CLIMATE CHANGE DENIAL

Peter Stott is a Science Fellow in Climate Attribution at the Met Office's Hadley Centre and Professor in Detection and Attribution at the University of Exeter. He has played a leading role in the United Nations Intergovernmental Panel on Climate Change and has been published in *Nature* and *Science*, among many other journals.

Hot Air

THE INSIDE STORY OF
THE BATTLE AGAINST
CLIMATE CHANGE DENIAL

Peter Stott

Atlantic Books
London

Published in hardback and trade paperback in Great Britain in 2021 by
Atlantic Books, an imprint of Atlantic Books Ltd.

This paperback edition published in 2022.

10 9 8 7 6 5 4 3 2 1

A CIP catalogue record for this book is available from the British Library.

Paperback ISBN: 978 1 83895 251 8
E-book ISBN: 978 1 83895 250 1

Printed in Great Britain by Clays Ltd, Elcograf S.p.A.

Atlantic Books
An imprint of Atlantic Books Ltd
Ormond House
26–27 Boswell Street
London
WC1N 3JZ

www.atlantic-books.co.uk

To Pierrette

Contents

Preface to the Paperback Edition

In early November 2021, three weeks after the first publication of *Hot Air*, I travelled to Glasgow to present the latest scientific evidence on climate change at the COP26 summit. With time running out to avoid catastrophic climate breakdown, these were the most important talks yet. As this book explains, the scientific case for rapid and substantial reductions in greenhouse gas emissions is indisputable. But, as I also recount here, for many years climate change deniers have gone to great lengths to prevent steps being taken to reduce those emissions. Their delaying tactics have held back progress despite a growing clamour from people worldwide for the crisis to be addressed before it is too late. Every year there are more and more extreme weather disasters. At COP26, policy makers needed to start putting a halt to the rising tide of human misery from fires, floods, droughts and storms.

With so much at stake, I was struck by the powerful appeal made by the prime minister of Barbados, Mia Amor Mottley, during the opening ceremony of the conference. 'We must act in the interests of all of our people who are depending on us,' she told fellow world leaders, 'and if we don't, we will allow the path of greed and

selfishness to sow the seeds of our common destruction.' Now or
never, she was saying, it was time for governments to signal their
firm intention to bring down emissions.

At none of the previous climate summits – many of which I have
attended since COP3 in Kyoto in 1997 – had the evidence of science
and the views of citizens been so much to the fore. For the first time
at COP, there was a science pavilion, a dedicated space within the
conference venue for scientists to present their findings to delegates.
I was responsible for drawing up the programme for the pavilion
in collaboration with colleagues from the World Meteorological
Organisation and the Intergovernmental Panel on Climate Change.
The evidence we had to show, including the recent report from
the IPCC published in August 2021, is stark and unequivocal.[1]
Widespread and rapid changes in the atmosphere and ocean, the
snow and ice, and the biosphere are already occurring. Carbon
dioxide emissions need to be reduced rapidly and reach net zero
to prevent global warming reaching catastrophic levels. Strong
reductions in other greenhouse gas emissions, such as methane, are
also required. The message is striking: no longer can policy makers
afford to ignore the urgency of that call to action.

Nor can they afford to ignore the will of their citizens. On 6
November 2021, the middle Saturday of the congress, I left the
science pavilion and the vast complex of exhibition halls and double-
decker tented structures surrounding it, and walked up the road
to where a steady stream of protestors, part of a 100,000 strong
crowd, were marching by in the rain. As I stood to cheer them on,
a dramatic rainbow appeared above their colourful banners. 'How
many COPs to arrest climate chaos?', asked one. It was an image
that seemed to symbolize the hope and anxiety swirling around
Scotland's most populous city as people there, and elsewhere around
the world, waited to hear the outcome of these vital negotiations.

The deal finally agreed to, the Glasgow Climate Pact, does not by
itself put the world on track to solve climate change. Much more is

yet to be agreed at future COPs and much more needs to be done by companies, local authorities and legislators. But the Glasgow Climate Pact does, potentially, mark a turning point in climate action by committing countries to substantial cuts in carbon dioxide emissions – a 45 per cent reduction globally by 2030 relative to 2010 levels – as well as deep reductions in other greenhouse gases.[2] Although worked-up plans by nations do not yet meet this level of ambition, governments agreed to come back each year with enhanced commitments for cutting their emissions. As the science and, increasingly, individual citizens demand, governments are accepting the need to accelerate their efforts.

There is no time to lose. Since I finished writing *Hot Air*, there have been numerous illustrations of the destructive power of extreme weather. During the summer of 2021, Europe's hottest on record, the continent saw its hottest ever temperature – 48.8 degrees Celsius in Syracuse, Sicily.[3] During that heatwave, devastating fires swept across Italy, Turkey and Greece destroying properties, crops and forestry. And fires are no longer a problem confined to the summer months. In Colorado, wildfires destroyed hundreds of homes at the start of 2022, a disturbing example of how such destruction is now afflicting people even in the middle of winter. During that time, even worse devastation was meted out to people living in parts of the world less able to cope than relatively prosperous Europe and North America. After facing two severe droughts and eight tropical storms over the previous five years, Mozambique was battered once again in January 2022, this time by Tropical Storm Ana. More than 10,000 homes were demolished as were powerlines, schools and health facilities.[4]

For many years, climate deniers have claimed that weather is not becoming more extreme. In the face of recent events, this claim is increasingly seen as derisible. But climate change denial has not gone away. The US could yet elect a president in 2024 who, like Trump, favours oil and gas, withdraws from the COP process and delays decarbonisation of the American economy. And in the UK, the emergence of a new group of Conservative parliamentarians provides a salutary reminder that this country's progress towards net zero is far from secure.

The Net Zero Scrutiny Group of MPs and members of the House of Lords blames soaring energy prices on the green agenda, seeks to delay action on climate change and supports further exploitation of fossil fuels. Its arguments are built on the same bedrock of climate change denial exposed in *Hot Air* that has done so much damage over the past three decades. The Global Warming Policy Foundation, one of the numerous lobby groups whose activities are also recounted in *Hot Air*, remains at the heart of this agenda of obfuscation and delay. The week after COP26 finished, their annual lecture for 2021 was delivered by American physicist Steven Koonin to an audience that included Steve Baker and Peter Lilley, both members of the Net Zero Scrutiny Group.[5] In it, Koonin claimed that the IPCC report had been distorted by alarmists, argued that authoritative bodies like the Royal Society and the US National Academies should declare that there was no climate crisis and insisted that as regards climate change: 'we can deal with it in due course – but let's all relax'.[6]

These are dangerous arguments. We can't relax: if we do, greenhouse gas emissions will continue to surge and humanity will face widespread drought and crop shortages, rising sea levels that make land uninhabitable for millions, and destructive storms that are impossible for communities to recover from. And if global warming continues, the great ice sheets of Antarctica and Greenland could reach a tipping point beyond which their melting is irreversible. Sea levels would rise 15 metres or more and there would be nothing that future generations could do to stop it.

Thankfully, such a terrifying prospect is not yet inevitable. By reaching the goal of net zero on the timetable agreed under the Glasgow Pact, this dystopian nightmare would most probably be avoided. Instead, climate action raises the much more attractive prospect of a resilient and sustainable future, one of renewable energy generation, nature preserved and communities able to cope with the vagaries of weather not distorted beyond the bounds of adaptability. Such a future won't come cost-free. But it is realistic and achievable according to the Climate Change Committee, the independent group of experts set up to advise the UK government. And it makes much more economic sense than trying to absorb the escalating costs of ever more extreme weather events that would be the inevitable result of failing to reduce emissions.

As this book shows, it has taken far too long for governments to treat the climate crisis with the seriousness it demands. But now that progress is finally being made, it is crucial that these early steps do not falter but, instead, gather pace as governments look to implement what was agreed in Glasgow. All the while, the dark forces of climate change denial remain in play, intent on obstructing any quickening of pace towards a greener world. In this light, the events recounted in *Hot Air* appear not only of historical interest, but startlingly relevant to today's situation. Developing the science, confronting denial and greening the economy remain the three vital pillars in our collective response to climate change.

The difference is that, now, time is running out. The longer humanity keeps emitting greenhouse gases into the atmosphere, the more difficult it becomes to steer the world away from the cliff edge of irreversible climate breakdown. And the Russian invasion of Ukraine has reminded democratic countries that their national security depends on weaning themselves off their reliance on oil and gas from totalitarian regimes. As Europe is convulsed by the horrors of indiscriminate bombing of civilians and millions fleeing in terror, so the imperative on governments has only become stronger to work together to preserve humanity from catastrophe – whether annihilation from nuclear war or unrestrained use of fossil fuels.

The benefits of overcoming the climate crisis have never been clearer. And while climate change denial remains, its false appeal shrivels under the scrutiny of citizens engaged in building a more hopeful future. With every step each of us takes towards a more sustainable lifestyle, the closer we come to a climate that supports the aspirations of future generations just as it has for generations before.

Peter Stott,
April 2022

Prologue

Approaching the closing moments of our annual scientific meeting I find myself getting emotional. Together with colleagues from the International Detection and Attribution Group, I've spent the last three days dissecting our latest research into the earth's changing climate. Our results are not reassuring: unprecedented heatwaves, more devastating storms, rapidly melting ice – all, we have found, are attributable to human-induced greenhouse gas emissions.[1] It's a sobering picture. But the research itself is not why I'm feeling increasingly distressed. There is still time – just – to avert global environmental catastrophe, as long as governments take our findings seriously. The problem is, I'm hearing, they're not.

One government in particular is, once again, disputing our science. It is the most powerful nation on earth, home to over 300 million people and our host country today. Two months ago, Donald Trump was inaugurated as US president, pledging to withdraw from the international Paris Agreement to tackle climate change, and installing Scott Pruitt as head of the country's Environmental Protection Agency. Pruitt, a man well known for working with oil and gas companies to challenge environmental regulations, rejects the scientific consensus on climate change.[2] According to him, carbon dioxide is not a primary contributor to climate change.[3] According to him, global warming has stopped.[4] Like the rest of my colleagues here, I am well aware that these are outrageous falsehoods.

For over twenty years, I have been a member of the International Detection and Attribution Group, a small band of twenty or so researchers from around the world whose quest has been to establish the causes of climate change.[5] It's this group that has proved beyond a shadow of a doubt that global warming is caused by human-induced emissions of carbon dioxide and other greenhouse gases. I've been at the forefront of that quest: I carried out the original climate model simulations that demonstrated how natural causes fail to explain recent warming and I was the first to link an extreme weather event – the devastating European heatwave of 2003 that killed over 70,000 people – directly to human activities.[6] I know how significant these findings – and those of my colleagues – are. That's why what I'm hearing today makes me feel so angry.

I'm ignoring the arresting view from our seminar room at the Berkeley Lab, a complex of multi-storey buildings perched on a steep hillside above the campus of the University of California, Berkeley: the glittering waters of San Francisco Bay, the Golden Gate Bridge, the rocky fortress of Alcatraz. Instead, I'm watching three large television screens arrayed across the front of the room on which our speaker, Ben Santer, is presenting illustrative slides. We could all be downtown by the Wharf, eating seafood in the warm sunshine, chatting about our families ahead of our travel home. But we're still here, in these closing moments of our annual gathering. We're keen to hear what he has to say, the colleague who more than twenty years ago kick-started this obscure, life-changing, bitterly controversial scientific field of ours. We want to know how Ben is fighting back against the lies being spread, once again, about the discoveries we've made, discoveries which affect all life on earth.

Ben is one of the founding members of our group, which was established in 1995. He has made a string of groundbreaking discoveries, including detecting the fingerprints of human activity in atmospheric temperatures, water vapour and ocean heat content. But today he is feeling too stressed to travel the forty or so miles from his office at the Lawrence Livermore National Laboratory to join us in person. Instead he is presenting his talk remotely, his voice coming to us, like a pained but determined spirit, from speakers hidden in the ceiling.

Ben knows how strong the evidence is for human-induced climate change. Over more than two decades, he has gone out of his way to explain to government officials, politicians and the media why we know that greenhouse gas emissions have caused global warming and why it matters. For his trouble, he has been accused of fraud, threatened with the sack, and charged with betraying his fellow citizens. He's been through tough times before. But now, after all this time, and as the consequences of further delay in tackling the climate crisis become ever more stark, it feels like this latest attack, by the man in charge of his country's environmental protection, is the worst of all.

Calmly, but with restrained fury, Ben tells us how he has gone about assessing Pruitt's claim that, over the last two decades, atmospheric temperatures measured from satellites have not increased. It fits the bill, Ben explains, for what is known in science as a testable hypothesis. It doesn't go against the scientific grain to make a supposition, no matter how outlandish. But for a supposition to have any validity as a scientific hypothesis, it must stand up to testing against observed reality. Although it's a test that Pruitt has no interest in carrying out, Ben has gone to the trouble of doing so himself.

Our colleague has calculated warming trends from all of the different satellite data sets available and worked out if any of them have dropped below levels that could be explained by natural variations in climate.[7] The results are quite clear: they haven't. In fact, Ben has shown that the chance of Pruitt's supposition being correct is vanishingly small. The well-established reality of global warming being a hoax is as unlikely as the experimentalists at CERN having made a mistake in detecting the Higgs boson. It's an analysis that Ben plans to publish in an academic journal. It won't change Pruitt's mind, but it is important to set it down, for the record.

What more can we do, I ask Ben after he has finished speaking, to get our message across? It is easy to make these misleading claims again and again is Ben's answer. We have to keep countering them, even if it's a slow process which gets much less attention than the original falsehoods. I can see his point. It's frustrating, but as scientists who care about our findings being taken seriously, it has to be done.

And yet, it doesn't seem enough. Rebuttals like Ben's reach only a small audience who have access to the relevant technical papers. A wider public, unversed in the details, can hear only claim and counterclaim, a false balance that can lead to too many people thinking too much is uncertain. The strength of evidence built up over years that points towards human not natural causes for global warming gets lost. It's a point that the forces of climate change denial know only too well. People like Pruitt, supported by a rump of pseudoscientific so-called 'climate sceptics', have weaponized doubt in the service of the fossil fuel industry. To preserve business-as-usual profits for as long as possible, their aim has been to promote delay in tackling this most urgent of global issues.

For a scientist who helped uncover that evidence and knows the terrible risks further emissions of greenhouse gases will bring, it is disheartening. Like Ben, I have been working in this field for over two decades, developing the understanding of climate change, explaining my findings and setting out what needs to be done to avoid the worst impacts of climate change. Like him, I have had to deal with attacks on my integrity. Like him, I had hoped that, after all these years of struggle, assaults like this would finally be over, that the overwhelming majority of governments and citizens would have accepted the reality of the climate crisis and have set about tackling it with the urgency it requires.

With these reflections comes a realization that there *is* something more I can do. There is a story I have to tell. It's one that has not yet been told but which needs to be widely understood if the planetary catastrophe caused by ongoing greenhouse gas emissions is going to be averted in the nick of time.

*

I have long been interested in the natural world. When I was growing up, I often went hiking in the Lake District with my family. It was there that I learnt to love the great outdoors with its precarious beauty and its possibilities for adventure. While at university, I worked during the vacations as a walks leader in Snowdonia, for which a keen appreciation of changes in the weather was as vital a skill as an ability to accurately read a map. With such a background, it was perhaps inevitable that one day I would combine my interests in science and the environment.

I studied for my PhD at Imperial College in London, where I researched the transport of radioactivity from the explosion at the Chernobyl nuclear reactor in 1986. I calculated how atmospheric winds carry radionuclides vast distances and how clouds concentrate them locally. My results explained why the sale of Lakeland sheep needed to be restricted for years after rain, polluted by an accidental release one week earlier and 2,000 kilometres away, had contaminated the fells.[8] The Chernobyl disaster provided a lesson to many across Europe: deadly atmospheric pollution is no respecter of national boundaries.

After my doctorate, I carried out research at Edinburgh University into ozone depletion. The ozone layer in the upper atmosphere protects people and animals from the damaging effects of ultraviolet solar radiation. But in the mid 1980s it was discovered that the ozone layer was being destroyed by chemicals released from fridges and deodorant sprays. Thankfully, in 1987, international action was taken to start replacing the offending chemicals with non-damaging alternatives. By the time my postdoctoral research contract came to an end in the mid 1990s, the concentrations of the offending chemicals in the atmosphere were starting to decrease. It would take many more decades for the ozone to recover, including the alarming ozone hole that had opened up above Antarctica. But more serious damage had been averted thanks to prompt international action informed by scientific advice.[9]

I was reluctant to leave Scotland. It was where my wife and I had met and where we enjoyed mountain climbing and exploring the West Coast's lochs and islands. But the Met Office was offering me a permanent job and Pierrette could transfer her music studies to Reading University. John Mitchell, the Met Office's chief climate modeller, recruited me to work on what was now the most pressing environmental issue facing humanity – climate change. I had been assigned to work in a fascinating new field of climate research, the search for the fingerprints of human-induced climate change in meteorological data. With a first degree in mathematics, I had long been intrigued by the presence of patterns in nature. Now I would be looking for unnatural patterns in nature, developing new mathematical techniques to hunt them down and using one of the world's largest supercomputers to do so.

I was delighted to be working at the Met Office's Hadley Centre for Climate Prediction and Research, one of the world's leading institutes in climate science.[10] But I quickly discovered that this bright new endeavour came with a darker side. The research was under attack from a plethora of lobby groups promoting climate change denial. This was not the normal cut and thrust of rigorous scientific critique, which is a legitimate and necessary part of the scientific process. This was something else entirely, a concerted attempt to discredit our work, not because of its shortcomings but because of its inconvenient implications for people whose vested interests could be damaged by what we found out.

I first met Ben Santer six months into my new job when I travelled to San Francisco for my first international conference in climate change. At an extraordinary session of the meeting, I saw him present groundbreaking new results linking atmospheric temperature changes to human-induced greenhouse gas emissions. I also saw his work attacked by a prominent 'climate sceptic'. This was a man who received funding from the fossil fuel industry and who had previously tried to discredit the science behind ozone depletion. What I found most shocking was his accusation that Ben had distorted the latest report of the Intergovernmental Panel on Climate Change (IPCC), the body mandated by the United Nations to assess the latest scientific understanding, supposedly to further his own political ends. It was a baseless accusation, promulgated in the media and the US Congress, and designed to nullify a leading scientist whose research had devastating implications: that climate change was already, by the mid 1990s, a significant threat to the lives, livelihood and prosperity of billions and one that needed to be tackled urgently. Attending that meeting in December 1996 gave me my first taste of climate change denial.

I would have many more: in December 1997 when I was confronted by a leading climate denier at the international climate negotiations in Kyoto where I had travelled to present our latest scientific findings; at an extraordinary show trial of climate science in Moscow in 2004 where the climate denier community gathered to support attempts to prevent Russia ratifying the Kyoto Protocol to limit greenhouse gas emissions; when I acted as an expert witness in the High Court during a case seeking to prevent the 2006 Al Gore documentary about global warming, *An Inconvenient Truth*, being shown in schools.

All this time, we members of the International Detection and Attribution Group were finding more and more evidence for human influence on climate. In January 2007, I was present at perhaps the most eagerly awaited and best attended press conference in the history of climate science when the IPCC published its newest report.[11] The scientific community had delivered their verdict – 'warming of the climate system is unequivocal' the report stated in the most strongly worded assessment yet – and soon it was the turn of governments to do their bit at a crunch climate summit in Copenhagen in December 2009. This was their chance to start turning our climate projections of ever more floods, droughts and famines, which were based on assumptions of continued emissions of greenhouse gases, into dystopian fictions. To do so they had to come to a collective agreement to drive down those damaging emissions.

The meeting ended in failure without agreement and in the wake of another vicious attack on science by the climate deniers. Alongside the failure of politicians, the 'Climategate' controversy, in which stolen emails were used to justify a false narrative that global warming was a hoax, damaged public trust in climate change as a pressing issue that needed to be urgently addressed. I had been working in the field for over a decade by then, and was leading a group of thirty scientists who were monitoring as well as attributing changes in climate. From what we were seeing – more intense heatwaves and floods, rapidly melting Arctic ice, rising sea levels – there was no time to waste in dealing with the mounting climate crisis. Yet time *was* being wasted, thanks in no small part, to the efforts of the climate deniers.

If there was hope, it seemed to come, ironically perhaps, from the very people most affected by the mounting toll of extreme weather events. In January 2013, having travelled to Tasmania for a meeting to develop the next IPCC report, I met a family who had narrowly escaped death when their home was suddenly engulfed by forest fires the week before. They, like the other people at a public meeting I addressed, knew that the record-breaking heat that led to the fires was no accident, that global warming was most definitely not a hoax. And they wanted to know what could be done to prevent more of such disasters in future. For many around the world facing the ravages of unprecedented weather extremes, reducing greenhouse gas emissions had become, literally, a matter of life and death.

Eventually, an international agreement to curb greenhouse gas emissions *was* reached, in Paris in 2015. Under the agreement, countries have started taking domestic action. In the UK, the Climate Change Act mandates government to reduce emissions in line with the latest scientific advice provided by the independent Climate Change Committee. To many, including those of us who found the evidence for human-induced climate change, it seems like there *is* hope, at long last, that the climate crisis can be solved, even though we have yet to see global emissions starting to reduce in practice.

Having spent twenty-five years on the front line of climate science, I know the stakes at play if humanity does not act quickly. Already, delay has incurred devastating costs in lives and livelihoods from increasingly damaging heatwaves, floods, droughts and storms. The costs of further delay are even greater. If global warming is allowed to continue unchecked, humanity might not be able to feed itself, many coastal cities could need to be abandoned, and large parts of the earth could become uninhabitable. To avoid the worst effects of climate change, countries pledged in Paris to keep global warming to well below 2 degrees Celsius relative to pre-industrial levels. To do so, emissions must start ramping down sharply over the next few years and reach net zero over the next three or four decades. The task is not impossible but it is now much harder than if humanity had acted sooner. For that, the climate change deniers bear a heavy responsibility.

I have a story to tell, an insider's guide to one of the most complicated and divisive issues of our time. It is a story to counter the false and damaging claims spread by people like Pruitt, to set the record straight. It is also a story of hope.

By eliminating our dependence on fossil fuels, we face a better future in which our children can breathe clean air and our grandchildren can enjoy the riches of a habitable and sustaining planet. The pedlars of doubt, with their vested interests in preserving the status quo, have persistently tried to deny the world that hope. It's a delay that has already cost us dearly, thanks to the rising impacts of a changing climate. If we want to reach that better future, it's a delay that can't go on.

The battle against climate change denial has taken place behind the scenes, in laboratories and conference halls, in courtrooms and parliamentary hearings. I know what went on. It's time that you did too.

1

Fingerprinting the climate

The week before Christmas 1996, I presented my first work on climate change at an international scientific meeting.[1] Usually, a debutant on the conference circuit gets a chance to break themselves in gently. They might present a talk at a low-key side meeting, put up a poster on a board surrounded by thousands of others, or simply listen to the findings of more experienced colleagues. Instead, I found myself describing my early results to one of the most eagerly awaited sessions that the annual meeting of the American Geophysical Union (an international non-profit scientific association) has ever seen.[2]

Each year, the Moscone Center, a cavernous underground complex of halls a couple of blocks from the heritage cable cars and thrusting skyscrapers of downtown San Francisco, welcomes tens of thousands of scientists from around the world. From the study of the sun's surface to the interior of the earth's core, from the weather on Mars to our own world's ocean currents, this huge meeting covers the latest advances in the geophysical sciences. It's quite a circus.

The wide subterranean lobbies bustle with activity. People queue up in front of coffee stations and hurry in and out of restrooms, old friends gather in sociable huddles and collaborators sit together at tables peering at laptops. In odd corners, smartly dressed young scientists, hopeful of impressing future employers, silently rehearse their upcoming presentations.

But it is through the double doors in the meeting rooms themselves that the real action of the week takes place. In a host of halls across the complex, postgraduate students, postdoctoral researchers, tenured academics and emeritus professors present their latest research findings from microphoned platforms to serried ranks of chairs. Delegates who are not presenting at that time can pick and choose from the multiple sessions taking place in parallel. They can dip in and out, grazing for the scientifically most interesting titbits, or settle into a session for the duration, spend the time with a community of specialists and catch up with all their latest progress. The scale of the centre and the grandeur of the rooms means there are usually enough seats for everyone.

Even before our session in December 1996 began, every spot was taken. The theme – the detection and attribution of climate change – had until recently been rather obscure and pursued by only a handful of researchers. But this year it had been thrust into the limelight. Right at the start of my career in climate science, I was going to be presenting my early results to a huge crowd. It felt like a very intimidating initiation into my new field of research.

At the start of the year, the Intergovernmental Panel on Climate Change (IPCC) had published its latest report.[3] Previously, the United Nations body charged with assessing the latest scientific understanding had not been able to say whether past warming was human-induced or natural in origin. Now, it had come to a very different conclusion, that 'the balance of evidence suggests a discernible human influence on global climate'. It was a conclusion that would have major repercussions.

A carefully considered summation of the latest findings by the small band of researchers who made up the International Detection and Attribution Group (which had been founded only the year before), the 'discernible human influence' statement prompted a bitter controversy. For the first time, the scientific consensus was that the finger of blame for climate change pointed firmly towards emissions of fossil fuels. This was not what the fossil fuel industry and many lawmakers in the US Congress wanted to hear. In recent months, they had tried to dispute the IPCC's findings in a controversy that had reached the pages of national newspapers and been aired in Congressional hearings. The two leading protagonists in that dispute were due to speak at today's session: Ben Santer, the climate scientist who had taken the leading role in crafting the 'discernible human influence' statement, and Patrick J. Michaels, State Climatologist of Virginia and vocal critic of the IPCC.

With that in mind, I shouldn't have been surprised to see so many people here. But it still came as a shock. For the first time it struck home that this research field of mine was of interest to an awful lot of people.

I had arrived in good time with my two colleagues from Britain: Simon Tett, my mentor from the Met Office Hadley Centre, and Myles Allen, his energetic collaborator from the University of Oxford, and we had found a place near the front next to an aisle. Even before the talks had begun, people had lined up against the back wall and along the huge sliding partitions that separated our hall from the much quieter ones on either side. Late arrivals were leaving again through the thick double doors behind us, looking disappointed. There was an excited hubbub of chatter and some of the other speakers were standing around at the front getting a feel for the atmosphere.

Despite the crowds, ours was a regular session of the conference that featured many technical presentations that provided incremental advances to scientific knowledge. My talk was going to present one of those advances. After just four months working on the subject of climate change, I could hardly have expected to have already made an earth-shattering discovery. But the work I was about to present was relevant to the high-stakes research question that had drawn in the crowds. I hoped people would find my findings interesting and worthwhile. Most of all, with so many eyes trained on me, I hoped I wouldn't make a fool of myself.

A hush had descended, the lights had dimmed, and the chair was inviting the first speaker up to the raised podium to begin the session. Not a word of what they said went in, nor did any of the colourful pictures projected on to the giant screen in front of me make any sense. Instead, I was consumed with apprehension as to how I would appear to all these hundreds of people, once I too had climbed the steps, clipped on my microphone and begun to talk. The time raced by. The audience clapped my predecessor. And then, without even being aware of how I had got there, I found myself standing on the same spot that the previous speaker had just vacated.

I dared to look down at the dimly lit sea of faces in front of me, took a deep breath and began to speak. Using data crunched by the Hadley Centre climate model, I told them, I had compared trends in surface temperatures with those expected from natural climate oscillations. Warming at the earth's surface, I had found, was now outside the envelope of temperatures that could be explained by natural processes.[4] To detect changes in climate, you had to look at the data over twenty years or more. Over a decade, temperatures could cool, even in the presence of greenhouse warming. With climate change it was important to look at the long-term picture. When you did, the data showed that recent warming was highly significant.

In a flash, my allotted time was up. People clapped politely and I was free to return to the safe anonymity of my seat in the vast crowd. A career hurdle overcome, I was at last capable of listening to the other speakers. Soon we would come to the main attraction. But first we would hear from other members of the supporting cast, including those of my slightly more experienced British colleagues, Simon and Myles. It was their boyish enthusiasm that had infected me with a strong passion for my newfound area of research. They had taught me a lot, these peers of mine, since I joined the Met Office Hadley Centre in July. They had been investigating the causes of climate change for over a year. I had several years' experience in other branches of atmospheric science, including researching the environmental consequences of the Chernobyl nuclear accident and the depletion of ozone by destructive chemicals. But I was new to this particular specialism and I had much still to learn. Compared with me, my two colleagues were veterans.

Simon rushed through his latest results with eager excitement. He had studied atmospheric temperatures high above the ground as measured over recent decades by weather balloons. He found that how they were changing could best be explained by taking account of human activities. His results confirmed previous expectations, that greenhouse gas emissions should warm climate, affecting temperatures not just at the surface but aloft as well.[5]

Myles was up next and looked totally at home in front of a large and expectant audience as he expounded his new idea. He wanted to improve our understanding of climate change by developing a new method for working out exactly how much warming was caused by human activities. Air pollution in past years had obscured the sky and shielded the earth from some of the sun's rays, holding back warming from the increasing greenhouse gases in the air. Exactly how much was still uncertain. Myles had developed a set of detailed equations to work it all out. His engaging style was attractive even though I couldn't see many people following his complicated mathematics.[6]

There were other speakers from other institutes. Like ours, they were technical addenda to the main business that had filled up this hall to overflowing. It was normal business during a technical session of a scientific congress. But unusually, rather than presenting to a handful of interested specialists, this time we had found ourselves presenting to the massed ranks of the world's geophysicists.

At last, we had arrived at the promised showdown when the two principals would be invited to make their respective cases. Patrick J. Michaels strode confidently around the stage, eschewing the mathematical equations and colourful illustrations that had featured in previous talks. Instead he talked to a sequence of bullet-pointed position statements projected on to the giant screen behind him. Global warming was not a problem, he claimed.[7] Natural processes caused much larger changes in climate than any human activities could produce. The IPCC had become politically compromised.

And then it got personal. According to Michaels, the work of Ben Santer was fundamentally flawed. Michaels claimed that the changes Ben had attributed to human activities in a paper recently published in the prestigious scientific journal *Nature* were in fact entirely natural. Temperatures in the lower part of the atmosphere had warmed differently in the southern hemisphere than the northern hemisphere, a feature that could not be explained by human causes according to Michaels.[8] Not just that, Ben had been instrumental in distorting the latest IPCC report, including its conclusion that there was 'a discernible human influence on global climate', for political ends. The effect of this political misuse of science, he claimed, was to deceive policymakers and the public into falsely believing human activities were causing global warming thereby promoting unnecessary restraints on economic growth that would destroy world economies.[9]

It was a slick presentation, easy to follow and clear in its conclusions. It was also now clear what all the fuss was about, why every seat was taken, and why an expectant stillness had fallen about the audience as we waited for Ben's response. Given the recent history of the climate change issue, I too was eager to hear what he was going to say.

The reality of the earth's greenhouse gas effect had been established back in the nineteenth century. And concentrations of carbon dioxide in the atmosphere, the main greenhouse gas associated with human-induced emissions, had risen steadily since monitoring began in the late 1950s. But widespread awareness of climate change as a global issue did not emerge until the late 1980s with a growing concern that atmospheric temperatures were also starting to rise. The United Nations established the Intergovernmental Panel on Climate Change in 1988 to assess scientific understanding of the issue. The same year, British prime minister Margaret Thatcher gave a speech at the Royal Society, in which she said that humanity had 'unwittingly begun a massive experiment with the system of this planet itself', and two years later she opened the new Hadley Centre for Climate Prediction and Research at the Met Office.[10] Then, in 1992, the United Nations held the Earth Summit in Rio de Janeiro at which nations agreed to stabilize greenhouse gas concentrations in the atmosphere 'at a level that would prevent dangerous anthropogenic (human-induced) interference with the climate system'.[11]

With the potential for global action on climate change, lobby groups started to form to fight back against possible regulation. Generously funded by the oil, auto and coal industries, groups like the Global Climate Coalition, founded in 1989 by ExxonMobil and the American Petroleum Institute, supported a small group of contrarian scientists – of which Michaels was one – to attack the science of climate change.[12] For many years, climate science had received little attention outside of research labs and specialist conferences. By the mid 1990s, all that had changed.

It was the emerging field of detection and attribution of climate change that was bearing the brunt of this new attention. In past years, the focus of climate research had been on trying to predict global warming by the end of the twenty-first century, a distant prospect of no immediate concern to many people. But now, thanks to the research carried out by Ben and other members of the International Detection and Attribution Group, the IPCC had concluded that climate had already, by the mid 1990s, changed significantly. The threat posed by emissions of greenhouse gases was no longer a distant one requiring action in years to come. Instead, we had shown, it was a present one, requiring action now.

The world was facing a huge decision, whether to cut back its use of fossil fuels and, in doing so, radically change its entire means of energy generation. That decision rested on the headline conclusion from the IPCC that humans were influencing climate. In claiming that this finding was fundamentally flawed, Michaels was disputing not just Ben's scientific insights and integrity but also the entire basis on which many nations were now arguing climate change should be urgently addressed.

Having met him a few days before at his lab nearby, the claim that Ben was part of an international conspiracy to fraudulently mislead the American public seemed ridiculously far-fetched. But this audience knew, like I did, that the scientific claims Ben was making were not beyond reasonable challenge. Identifying the signal of human-induced climate change amid the large natural variations of our constantly varying climate pushed at the limits of current climate science. Observations were limited and climate models were still in the early stages of development. The crowd of sceptical geophysicists here in the hall were not going to accept claims about the causes of climate change, whichever way they went, without convincing evidence.

Ben's presentation style was very different from Michaels. Rather than roam about the stage like Michaels had done, Ben sheltered behind the lectern while he earnestly and laboriously laid out his case by means of a sequence of detailed illustrated slides. He first explained how atmospheric temperatures had altered over the last twenty-five years, as observed by weather balloons. Pointing to a large coloured map showing how temperatures had changed with height and across the latitudes of the earth, Ben showed us that there had been dramatic cooling high above the South Pole. Cooling was also present elsewhere in this upper part of the atmosphere called the stratosphere, the atmospheric layer where ozone shields the earth's surface from the harmful effects of ultraviolet radiation. Lower down, in the layer of the atmosphere which contains our weather systems called the troposphere, there had been warming. But just as Michaels had already described, this warming was not uniform. Instead, it was greater in the southern hemisphere than the northern hemisphere.[13]

To explain this complex picture properly, Ben told us, you had to consider how all possible climatic factors had affected the atmosphere. To illustrate this, he showed us some more plots, this time of how temperatures responded individually to the different factors, as calculated by the climate model. Ozone-destroying chemicals had cooled temperatures in the stratosphere. Burning fossil fuels had increased the concentrations of carbon dioxide and other greenhouse gases in the atmosphere, warming the troposphere and further cooling the stratosphere. And burning fossil fuels had another important effect.[14]

Small particulates of pollutant and water, known as aerosols, had formed in the air. These aerosols had reflected sunlight and made clouds brighter. Their effect was to cool climate. The importance of this effect had been demonstrated by my boss John Mitchell and other Hadley Centre colleagues including Simon Tett in a paper published the previous year in *Nature*.[15] By including this effect in the Hadley Centre climate model, they had improved the model's simulation of past temperature changes at the surface. It was evidence that global warming from increasing greenhouse gas concentrations was being moderated slightly by the cooling effects of aerosols.

Ben had looked into the effects of aerosols in more detail. Because there was more industry in the northern hemisphere polluting the air, the cooling effect of these pollution aerosols was greater north of the equator and the overall warming was consequently less there than in the south. Michaels had claimed the different rates of warming in the two hemispheres could only be explained by natural factors. Instead, Ben concluded, human-induced pollution was a much more likely explanation.

Crucially, the troposphere had warmed in both hemispheres, and this warming could only be explained by the increasing concentrations of greenhouse gases in the atmosphere. Thanks to a combination of ozone depletion and increasing greenhouse gases, the stratosphere had cooled. The observed pattern of tropospheric warming and stratospheric cooling was a smoking gun that pointed firmly towards human not natural causes.

Ben had investigated all the possible alternatives and come to the only possible conclusion given his careful examination of all the available evidence. Humanity had altered the planetary atmosphere in a significant way. If correct, and his case was persuasive, there was no going back from this. To understand more about how emissions were altering earth's climate, we scientists would have to seek out the fingerprints of human activity in other aspects of climate, including in temperatures across the surface of the oceans and land, in rain and snowfall, in droughts and storms. And as our evidence grew, governments and citizens would need to take our findings seriously before it was too late to save a planet's climate that over past millennia had given civilizations the opportunity to flourish.

Emerging blinking into the warm West Coast sunshine, Simon, Myles and I went in search of lunch. At a nearby café we bought ourselves a sandwich and a soda. Fired up by the session, Simon and Myles embarked on a vigorous discussion about how we could improve our techniques for calculating exactly how much of past warming was due to greenhouse gas emissions. Pleased that I'd survived my first talk at an international climate conference, I just wanted to eat.

Later I went shopping. With Christmas fast approaching I had presents to buy. Now that the main purpose of my trip had been accomplished I could allow myself to break away from my fellow British detectors and walk downtown towards the decorated tree in Union Square and Macy's department store. Cable cars clanked by, shoppers thronged the sidewalk and I thought about the conference.

Michaels' accusation that Ben had manipulated the report by the IPCC seemed outrageous. Ben had given a large part of his life over the last three years to the IPCC, working alongside many other scientists in carefully and sceptically assessing all the available evidence. Yet he had come under vicious attack from Michaels for doing so based on a partial and misleading interpretation of reality. Michaels may have had the greater oratorical skill, but Ben's scientific profile was much higher. That didn't seem surprising, given their different attitudes to winkling out the truth behind events. Michaels and his fellow contrarians had been dubbed climate change sceptics. But the true sceptics were scientists like Ben who based their conclusions on findings that stood up to critical scrutiny by fellow experts. Climate change denial seemed a better description of what Michaels was up to.

Now people had seen for themselves how flimsy the scientific arguments of Michaels were, surely they would dismiss his depiction of an IPCC report-writing process he had not participated in himself and few people outside the small group of scientists involved properly understood. More widely, people should assess the scientific evidence for climate change on its merits, not on rhetorical descriptions no matter how colourful. As the climate change issue rose in prominence, it would become more and more important that the mainstream scientific view was taken seriously and not derailed by a few contrarian voices. Surely, progress on tackling climate change would benefit from well-founded scientific advice, and not be hindered by partisan lobby groups well funded by the fossil fuel industry.

Maybe it was the balmy California weather that made me feel so optimistic that rational argument would prevail. Winter seems a long way off when temperatures are nudging 70 degrees Fahrenheit and there is a clear blue sky overhead. If reasonable people no longer doubted there was a link between greenhouse gas emissions and global warming, now it would be all about figuring out how bad the climate change problem really was. Perhaps Ben and the rest of our quirky little battalion of climate investigators could get on with the job, free of political interference and intimidation?

To my surprise, this optimism endured my flight back across the Atlantic. Sitting at my desk at the Hadley Centre, an anonymous three-storey office block on a dual carriageway in the suburbs of Bracknell, I enthusiastically set about writing up the results I'd presented in San Francisco. Working with my mentor, Simon Tett, I prepared a draft paper for submission to an academic journal. Our paper, which we titled, 'Scale-dependent detection of climate change', showed that global warming would take at least two decades to manifest itself. This is what we had seen over the previous twenty years as significant global warming emerged. Equally though, a temporary hiatus in warming was possible, even with substantial warming from greenhouse gas emissions.

It was exciting to submit my first scientific paper to the highly respected *Journal of Climate* such a short time after starting work at the Hadley Centre.[16] Having trained as a mathematician for my first degree, this opportunity to bring my analytical skills to bear on the emerging issue of climate change was one I grasped eagerly. Even so, I was joining a branch of the science involved – detection and attribution – that hardly anyone in the UK appeared to have heard of. Only when I found myself in the crowded and expectant hall at the American Geophysical Union in San Francisco had I realized what I'd let myself in for, and how significant this new field of research would turn out to be.

And now I had a result that was ready for publication. If it wasn't revolutionary, my finding did add another piece of the puzzle of what was going on with global warming. If after being reviewed by experts my paper were accepted by the journal, this would be my first peer-reviewed publication in climate science. I would have joined the lineage of those who had advanced our collective understanding of climate change, taking it from a fringe issue understood by few to a story that would appear on the front pages of national newspapers around the world day in day out.

*

That lineage began with French mathematician, Joseph Fourier, who in the 1820s proposed that the earth's atmosphere was warmer than it would otherwise have been thanks to what is now known as the 'greenhouse effect' (although Fourier did not use that term himself, the first usage of the greenhouse analogy being attributed to a Swedish meteorologist called Nils Ekholm in 1901).[17] The confirmation of the greenhouse effect has traditionally been credited to Irish scientist John Tyndall who, in 1859, experimented by shining light through a variety of gases and found that carbon dioxide soaked up energy in the infrared part of the spectrum. But a recently digitized copy of the *American Journal of Science and Arts* proves that a similar discovery was made three years earlier by a female scientist called Eunice Foote.[18]

Eunice Foote was a scientific researcher, inventor and women's rights campaigner who lived in New York State and whose experiments involved exposing tubes containing different gases including carbon dioxide to sunlight. Her results were presented to the Annual Meeting of the American Association for the Advancement of Science in 1856, not by her but by a male scientist called Joseph Henry (it being uncommon at that time for women to speak at such meetings) and were then published in the *American Journal of Science and Arts*. 'The highest effects of the sun's rays I have found to be in carbonic acid gas,' she wrote, carbonic acid gas being the term used at that time for carbon dioxide. 'An atmosphere of that gas', she continued, 'would give to our earth a high temperature; and if as some suppose, at one period of its history the air had mixed with it a larger proportion than at present, an increased temperature... must have necessarily resulted.'

Foote did not limit the radiation shining through her apparatus to the infrared part of the spectrum as Tyndall, who had access to more sophisticated equipment, would do later.[19] But she *was* the first to make the connection between carbon dioxide and climate change. When Tyndall made the same connection later in publishing his findings, he did not appear to be aware of Foote's work. In any event, by the 1860s the demonstration had been made that atmospheric carbon dioxide was a powerful absorber of infrared radiation – heat, in other words. Incoming energy from sunlight warms the earth's surface, which then radiates energy upwards at longer infrared wavelengths. This infrared radiation excites the molecules of carbon dioxide, methane and water vapour in the air, and warms the atmosphere. Without these 'greenhouse gases' in our atmosphere, solar energy absorbed at the earth's surface would be radiated straight back to space and it would be about 33 degrees Celsius colder as a global average. Instead, like a well-equipped climber wrapped up in layers of woolly fleece, greenhouse gases keep more of our planet's body heat in, allowing life to flourish.[20]

It was a Swedish chemist called Svante Arrhenius who, in the 1890s, first looked in more detail at what would happen if the concentration of greenhouse gases in the atmosphere changed. He made a rough calculation of the impact of doubling carbon dioxide, taking account also of the resultant increase of atmospheric water vapour, another greenhouse gas being evaporated off a warmer ocean. His answer was that the earth's surface would be about 5 degrees Celsius hotter, a result not that different from modern estimates. Thanks to Fourier, Foote, Tyndall and Arrhenius, the groundwork was laid for the now long-accepted fact of greenhouse warming.

What was not long accepted was whether human-induced emissions of greenhouse gases, including carbon dioxide, methane and nitrous oxide, could change climate enough to matter. In 1938, Guy Stewart Callendar, a British engineer and amateur meteorologist, proposed that global temperatures were rising due to increasing carbon dioxide levels, although most scientists, including Callendar himself, were sceptical this was a significant effect.[21] The question remained: was the artificially enhanced greenhouse effect of human activities an insignificant perturbation to a naturally variable climate that had seen Vikings growing barley in Greenland and Londoners throwing Frost Fairs on the Thames or was it capable of taking our planet into a vastly different and deeply perilous climatic state?[22]

The first step to answering this question was made by Charles David Keeling who set up an observatory in 1958 at the summit of the Mauna Loa volcano in Hawaii to monitor carbon dioxide levels in the atmosphere. These remarkable measurements taken ever since by him and his son Ralph Keeling have shown a steady rise of this powerful greenhouse gas. When they started, more than sixty years ago, concentrations measured less than 320 parts of carbon dioxide per million parts of air. Ever since, concentrations have been ticking upwards, to start with by about one part per million each year and more recently by over two parts per million each year, at a rate such that carbon dioxide concentrations have now well surpassed 400 parts per million.[23]

Nine years into the Keelings' remarkable testimony of rising carbon dioxide levels, Japanese scientist Syukuro Manabe, who was based at the Geophysical Fluid Dynamics Laboratory in the United States, published the results of an early climate model with fellow researcher Richard Wetherald. The Manabe and Wetherald paper is on the obligatory reading list of any aspirant climate scientist because it announces a crucial discovery.[24] The incoming radiation from the sun and the greenhouse effect are not the only factors affecting atmospheric temperatures.

The other vital factor, Manabe and Wetherald discovered, is the process by which temperature changes at the surface are closely tied to temperature changes above. When moisture is evaporated from the surface it cools the surface and warms the atmosphere when that moisture condenses to form water droplets in clouds. Their climate model was the first to take this process into account. As a result, it was the first climate model to provide a realistic representation of the vertical pattern of temperature changes as greenhouse gases increase.

Throughout the 1970s and 1980s, the Keeling measurements showed atmospheric carbon dioxide concentrations continuing to rise inexorably upwards. This was also a period when surface temperatures started increasing after a period of stagnation in the 1950s and '60s.[25] Up until then, the link between greenhouse gas concentrations and global warming had been a largely theoretical one, a question of what might happen in future based on physical understanding and early climate models. Now, people were starting to ask whether this link had become a practical reality. Were the higher carbon dioxide concentrations being measured in Hawaii related to the global warming now being observed at weather stations across the planet?

In dramatic testimony to the US Senate on a swelteringly hot day in Washington DC in June 1988, NASA scientist James Hansen declared that global warming from greenhouse gas emissions 'is already happening now'.[26] In charge of monitoring global temperatures at

NASA's Goddard Institute for Space Studies, Hansen's bold statement prompted strong responses from politics and industry.

The Republican candidate for that year's presidential election, George H.W. Bush, promised to address climate change by countering the 'greenhouse effect with the White House effect'. The oil industry too woke up to the potential threat of Hansen's statement, in their case not of potential disruption to the climate but of possible disruption to their profits from regulation of their activities. They began to fund lobby groups to cast doubt on the scientific basis for any such action. One such group, the George C. Marshall Institute, issued its first report attacking climate science the following year and followed it up with a briefing in the White House. Their activities helped to stop the incoming Bush administration taking the global warming issue seriously.[27]

Hansen's statement has since stood the test of time. But back then, there was no definitive proof that atmospheric temperatures really were changing unnaturally. Without such proof and with little political will to try to reduce mounting greenhouse gas emissions, the search was on to find the incriminating fingerprints of human interference on climate.

It was Klaus Hasselmann, the charismatic director of the Max Planck Institute for Meteorology in Hamburg, who initiated this scientific quest. Early in his tenure leading this prestigious institute in the late 1970s, Hasselmann's main interest was in understanding the basic workings of the earth's oceans. He was drawn to the properties of waves on the sea surface and the behaviour of water in the dense, dark depths below the surface.[28]

To help him, this wide-ranging intellectual heavyweight drew from the forefront of physical and mathematical theories including quantum mechanics, plasma physics and statistical science. He developed a new technique for detecting geophysical signals in noisy data which could be used to winkle out systematic changes in climate from natural variations.[29] Such variations include the El Niño–Southern Oscillation,

a huge swing in temperatures of the Pacific Ocean that brings warmer
and colder surface waters to the western coasts of North and South
America, and the Atlantic Multidecadal Oscillation, a decades-long
variation in the current bringing warm salty water from the Gulf of
Mexico up to the high Arctic. Hasselmann's institute had made many
advances in understanding these climatic oscillations. But eventually
the question of whether climate was changing due to human activities
became too compelling for him to resist.

In 1987, Hasselmann decided to apply his technique for detecting
signals in noisy data to the problem of climate change. He hired a young
postdoctoral researcher to tackle the problem, an American citizen
who had grown up in Germany and who had recently completed a
PhD at the University of East Anglia. His name was Ben Santer, and
that appointment would mark the beginning of the field of study that
became known as the detection and attribution of climate change.

Ben was asked to figure out what had caused the rise in temperatures
at the surface over the past century. Global warming had not been
steady but had risen most rapidly during the three decades leading up
to the Second World War and then had surged again from the 1970s
onwards after a prolonged period during the 1950s and '60s when
temperatures had stayed largely constant. Warming was generally
greater over land than oceans. The Arctic had warmed faster than
anywhere else. It was a complex picture. To decode it, Ben used
Hasselmann's detection technique for finding signals in noisy data.
And he used the results of climate models.

Climate models have provided the basis for many of the
discoveries made in climate science over the past few decades.[30]
A climate model is a bit like a film set. From the outside it's nothing
special, just many thousands of lines of impenetrable-seeming
computer code. Like the unglamorous sheds at Pinewood Studios,
it's hard to imagine that anything remarkable could happen here.
But once inside you can see the magic. Like the miracle of lights,
camera, action, so a climate model recreates a whole world of

movement. The winds blow, the oceans roar, the seasons come and go, the ice melts and freezes. And with such a tool Ben could explore our world, find out what would happen if the concentration of greenhouse gases were to change, the output from the sun vary or air pollution cause the release of climate-cooling aerosols.

Ben also needed a climate model to simulate how climate could vary naturally by itself. Natural climate oscillations, like the El Niño–Southern Oscillation or the Atlantic Multidecadal Oscillation, are emergent features of climate models, the physical equations of motion described in thousands of lines of computer code causing atmospheric temperatures to rise and fall periodically for years and decades at a time. These naturally occurring oscillations could explain at least part of the observed warming. Increases in solar activity could also have caused part of it. Air pollution could have cooled climate, particularly in the middle of the century, the time of the great smogs in the big industrial cities. And greenhouse gas emissions, from burning coal and other fossil fuels, could also have been a significant factor.

Ben had a lot to do and he needed help. He contacted colleagues from the Max Planck Institute and other climate centres who could assist with incorporating each possible effect into the climate models. He gathered up reams of data, not just from the models but from weather observations taken from far and wide. Then he compared his models with reality. By using Hasselmann's detection technique, Ben hoped to distinguish the systematic surges in temperature caused by the human-induced drivers of climate, like increasing greenhouse gas concentrations, from the natural up-swings and down-swings in temperature caused by, for example, solar variations or the El Niño–Southern Oscillation. He received some guidance from his heavily committed directorial boss. But ultimately, it was down to Ben to piece together the pieces of this climate puzzle. The earth's climate is a complex beast and Ben had a lot to wrestle with.[31]

The Max Planck Institute operates an arbitrary but strictly enforced policy that postdoctoral researchers are allowed to stay for no more than five years. By 1992, Ben's time was up. No matter that he was an outstanding young scientist making steady headway on a difficult problem. He found himself looking for a new job and Hasselmann found himself looking for a new recruit.

Ben's successor was a mathematically gifted researcher who had recently graduated with a doctorate in numerical fluid dynamics from the Ludwig-Maximilians University in Munich.[32] Gabi Hegerl was as determined to make her mark in this new field of research as her predecessor. Picking up where Ben had left off, Gabi continued the quest to identify the causes of temperature changes at the earth's surface. She realized that the effects of human influence on climate should be getting greater with time, increasing the level of agreement between the observed pattern of warming and the expected fingerprint of greenhouse gas influence. In earlier years, when the signal of human-induced climate change was rather small relative to the noise of natural climate variations, any agreement would be insignificant. But that could since have changed. Gabi wanted to see if the rising level of agreement between the greenhouse gas fingerprint and observations had now become significant.

Ben was recruited to work at a climate modelling institute situated on the California plains forty miles east of San Francisco that formed part of the Lawrence Livermore National Laboratory.[33] The new employment made him rethink the focus of his research. As a result, he started to think about the thermal structure of the atmosphere.

Above the surface, temperatures drop by about 6 degrees Celsius for every 1,000 metres gain in altitude.[34] The air gets thinner with height as less air bears down from above. At 4,000 metres, the altitude of many Alpine peaks, pressure has dropped by about 40 per cent and there is already a need for acclimatization, time for the body to adjust to the poorer levels of oxygen it finds in every breath.

All the mountains and all the weather on earth are found in the troposphere, the lowest part of our planet's atmosphere. It is a remarkably thin layer, only 18 kilometres high at the equator and

6 kilometres at the poles. If the earth were the size of a wooden desk globe, the troposphere would be little more than the thickness of two layers of varnish.[35] As air rises, it expands and the air molecules vibrate less energetically, which leads to cooling at the rate of about 10 degrees Celsius per 1,000 metres. But in many places moisture in the air condenses to form clouds, and this condensation process releases heat as water makes the phase transition from its vapour into its liquid form. Now the cooling is moderated by the latent heat of condensation. The net result is that, on average, air cools with height at a rate known as the moist adiabatic lapse rate, about 6 degrees Celsius per 1,000 metres.

Above the troposphere, in the next layer of the atmosphere called the stratosphere, temperatures stop falling with height and start to rise. Ozone is the cause. Energized by solar radiation, the same molecule that protects us from skin cancer by absorbing the sun's harmful ultraviolet rays also warms the air. Considering the troposphere and stratosphere together, atmospheric temperatures have a distinctive profile, cooling with height for the first 6 to 18 kilometres depending on whether you're nearer the poles or the equator, until you reach the tropopause, the point at which temperatures stop falling with height and begin to rise.[36]

Ben realized that the difference between how the troposphere and the stratosphere reacted to greenhouse gas emissions would be a distinctive fingerprint of human-induced climate change. While the troposphere warms in response to increasing greenhouse gas concentrations, the stratosphere cools. This is because the outgoing energy the earth radiates out to space has to equal the energy coming in from the sun. As the surface and lower atmosphere warm up due to the enhanced greenhouse effect, the upper atmosphere has to cool down to compensate, thereby keeping the overall energy radiating to space the same.[37] If the sun were responsible for global warming, the atmosphere wouldn't do this. Instead there would be warming at all heights, not just in the troposphere.

Klaus Hasselmann's two protégés were leading the way in the hunt for climate signals. A few others were also involved in the search. In 1994, Australian scientist David Karoly found a significant pattern of tropospheric warming and stratospheric cooling in data from weather balloons, and in 1995, my colleagues at the Hadley Centre, including John Mitchell and Simon Tett, found a significant pattern of warming at the surface.[38] But most of the efforts in climate science were in other aspects of the problem. Scientists were working hard to improve the representation of clouds, ice and oceans in climate models at centres in Australia, Canada, Germany, the United States and the UK where the Hadley Centre was gaining a leading reputation for climate research. And with those climate models, they were seeking to map out the potential changes in temperature, rainfall and sea level if greenhouse gas emissions continued unabated. There was a lot more to do if individual countries were going to be equipped with detailed projections of how their climate was set to change in coming decades. All the while, the definitive evidence that would prove climate had already changed due to human activities, remained elusive.

All scientific research is a race, a race to complete analyses before the funding dries up and to stay ahead of the scientific pack. Gabi and Ben were also in a race against a clock ticking down to publication of the latest report from the IPCC. Oil-producing countries like Saudi Arabia and Kuwait were quite happy that the previous report of the IPCC published in 1990 had been unable to say whether global warming was due to natural or human factors. They did not want to see blame levelled anywhere, least of all at burning fossil fuels. But for this second report of the IPCC, due to be finalized in late 1995, most countries were keen to see a more definitive conclusion if possible.

Gabi investigated temperature changes at the surface while Ben investigated temperature changes in the troposphere and stratosphere. They both faced the time-consuming task of compiling information from measurements made over many decades and comparing them with information from climate models. They analysed thousands of years of climate model data and millions of data points. Gabi worked with the University of East Anglia where researchers had compiled weather station data and ocean temperatures taken from ships and buoys. Ben contacted the Geophysical Fluid Dynamics Laboratory, which had compiled a data set of weather balloon measurements of atmospheric temperatures. It was laborious work. But it was necessary if they were going to find convincing evidence that pointed to the true cause of past temperature changes.

At last, just in time for their results to be included in the new report, the two pioneers in the detection and attribution of climate change published their groundbreaking findings. Gabi had found there was only a small chance that warming at the surface could be accounted for by natural factors. Ben had concluded there was little chance that a natural explanation could account for temperature changes aloft. Taken together, these results provided the most compelling case yet seen that human greenhouse gas emissions bore much of the blame for recent warming.[39]

The IPCC report was finalized at a meeting held in Madrid in November 1995. It was here that a short summary document representing the main conclusions of the report was agreed word by word by leading climate scientists and government representatives. Ben had led the writing of the chapter in the underlying report dealing with the causes of climate change. It was in this role that he went to the Madrid meeting, so he could help to ensure that the summary document correctly represented the scientific understanding of the causes of climate change.

The debate about how best to summarize the role of human influence on past warming took up much of that meeting. In the end, it all came down to one crucial sentence and took many hours – battling with the Saudi and Kuwaiti delegates who wanted to play down the scientific confidence in recent findings – before it was agreed. The resistance to scientific findings from these oil-producing countries in Madrid provided a foretaste of what was to face many more climate scientists in the years to come. But on this occasion, just a few minutes before the meeting was due to finish, governments unanimously endorsed the statement: 'The balance of evidence suggests a discernible human influence on global climate'.[40]

This carefully constructed set of words took account of the remaining scientific uncertainties involved. Climate models were imperfect and observed changes had to be worked out from incomplete and imperfect measurements. Nevertheless, there was convincing evidence that humanity's fingerprints were present in observed temperature changes. As a result, the world's governments now accepted that global warming could be linked to human rather than natural causes.

The 'discernible human influence' statement was a landmark conclusion that reverberated around the planet. Countries now had a clear imperative to act on climate change. Thanks to the march of scientific progress, from Joseph Fourier who discovered the earth's greenhouse effect, to Gabi Hegerl and Ben Santer who identified humanity's fingerprints on rising global temperatures, governments knew enough to start agreeing binding targets for reducing their emissions.

They would still need to work out exactly what these targets would be. Given the dependence of many economies on fossil fuels, this promised to be a difficult and messy challenge for politicians to address. They would have to figure out which countries should reduce their emissions most, and how quickly they should aim to do so. But at least they had agreed on the science as presented to them by the IPCC. Human activities were causing the world to warm.

Just as politicians prepared to act, so the backlash against scientific progress began. Ben Santer had been at the forefront of that progress, writing one of the seminal research papers and leading the drafting of the IPCC report. As a result, he was the scientist most exposed to attack and would come to feel the wrath of forces opposed to any change in the fossil-fuel-powered status quo.

In the years to come, many others would come under fire including myself. The climate deniers had only just begun their attempts to derail progress in dealing with climate change. The antithesis of true sceptics, they would spread confusion by cherry-picking data, targeting individuals and promoting unsubstantiated claims. The stronger the scientific consensus became, the harder they would work to frustrate attempts by the mainstream – and genuinely sceptical – scientific community to warn the world about the dangers of continued greenhouse gas emissions. We climate scientists would need to convince governments and citizens of the reality of those dangers, a message that they might not want to hear. It wasn't going to be easy. And climate deniers would do their utmost to make it a whole lot harder.

With my presentation at that highly charged session in San Francisco, and the acceptance of my first climate science paper in a peer-reviewed journal the following year, I was now in their sights. I was a published researcher and a member of the small group of climate change detectors who were in the vanguard of the developing science of climate change. At the end of the year, governments would meet in Kyoto to try to thrash out a deal to reduce their emissions of greenhouse gases. I was asked to travel to the meeting in Japan to present our latest results. I would not have to wait long to come face to face with one of the leading lights of climate change denial.

2

Confronted by denial

In December 1997, the British government asked the Hadley Centre to attend a meeting in Japan where countries hoped to thrash out a deal on climate that would start to reduce global greenhouse gas emissions. It was the most important international gathering on climate yet held. My boss, John Mitchell, selected two of the more junior of his team of researchers for the mission of informing the delegations about the science: Cath Senior, a physicist working on predicting future climate, and me.

Cath and I were both greenhorns at dealing with politicians, diplomats and climate sceptics – it was only my second year as a climate scientist – but we were eager to have a go. The historic city of Kyoto is now known for its role in setting a landmark in climate policy. For the first time, governments agreed to reduce their emissions of warming greenhouse gases. Without the Kyoto Protocol, there would have been no Paris Agreement, the global accord signed in 2015 that commits nations to keeping warming at sustainable levels and which now provides a beacon of hope that unsustainable climatic disaster can be averted. But back in 1997, there seemed little hope at all. Cath and I knew that global warming posed a massive problem to the well-being of present and future generations. Action was needed. But bureaucracy, pride and deliberate muddying of the waters are powerful agents for the status quo. In fact, what we witnessed in Japan's old imperial capital made it clear just what we climate scientists were up against.

In 1992, under the United Nation's Framework Convention on Climate Change, 154 nation states had pledged to reduce their emissions of greenhouse gases in such a way as to stabilize their atmospheric concentrations and prevent dangerous human interference with the climate system. The UN convention had been informed by the Intergovernmental Panel on Climate Change (IPCC), whose first report from 1990 had highlighted the disturbing rise in greenhouse gas concentrations in the atmosphere. But the 1992 convention had only asked for *voluntary* efforts to reduce emissions. Now, five years on, it had become clear that these voluntary efforts were not working. Concentrations of carbon dioxide and other greenhouse gases were rising as fast as ever. What was needed in Kyoto was a specific target for each country to reduce its emissions within a specified time frame.[1]

Agreeing such specific targets was a huge ask. Even if most states accepted the latest IPCC conclusion, that there was 'a discernible human influence on global climate', signing up to emission reductions would be an unprecedented step. Countries would have to accept international controls on their sovereign activities and they wouldn't commit to that unless they were convinced that the proposed reductions would be fair and equitable. And they needed to be sure that these major changes were worth it, that the science really did support such drastic action.

Cath and I had done our homework, but we knew there was a gargantuan task ahead. We had set up our Hadley Centre stand in a strategic position just a few steps from the main hall and next to a busy side room where a steady stream of delegations came and went. Their discussions proceeded all day and late into the night, and our shifts were just as long. We had climate projections to present: melting ice, rising sea levels and changing rainfall patterns. And we had comparisons to show, between the warming we'd already observed and the temperature changes expected from human and natural causes. Because we were the only climate modelling centre present, the appetite of delegates to hear what we had to say was huge. We knew that the Hadley Centre had placed a significant burden on our young shoulders. We did not want to let our bosses down.

We took comfort in developing a familiar routine in this most unfamiliar of situations. At breakfast, we navigated a hotel spread of fermented beans, miso soup and raw fish before queuing up at the nearby metro station for our train across town, waiting obediently behind the white lines which separated embarking from disembarking passengers before rattling across the city in carriages crowded with passengers who kept heads bowed in deference to fellow commuters. As the days followed on, we got to recognize the protesters against environmental destruction camped outside the modernist conference centre with its maple tree gardens, and the blue-shirted UN guards who supervised our passage through the airport-style screening stations into the lobbies beyond. Each morning before the crowds arrived, we eased ourselves into the day ahead by fetching over cups of coffee from the concession stand beside the lifts, booting up our bulky desktop computer and laying out a new pile of brochures.

The pace soon picked up as the halls filled with delegates. Visitors to our stand could be a politician or a civil servant, a scientific novice or a national expert, an industry lobbyist or a climate sceptic. Our job was to give them a copy of our brochure, a sixteen-page report on climate change and its impacts, and explain its contents.

The findings we had to present were novel and still controversial. Temperatures were rising, and 1997 was set to be the warmest year in a record stretching back to 1860. Changing atmospheric temperatures were fingerprints of human-induced climate change, not natural climate variability. Our climate models predicted a warming of about 3 degrees Celsius over the next century if greenhouse gas emissions continued unchecked. They projected increased hunger for millions, water shortages for those living in the subtropics and flooding for inhabitants of the world's coastlines. It was a frightening picture and one to be avoided if at all possible.[2]

Conversations started hesitantly as I tried to gauge a visitor's interests and understanding. The director of an African Meteorological Service, here to provide technical advice to his nation's delegation, greeted our results with relish. I had provided him with further ammunition in his argument with the developed world who needed to take their responsibilities seriously and cut their emissions. A policymaker, who may have had plenty of political and financial expertise but appeared to have little understanding of basic physics, needed me to explain the basics of the greenhouse effect. A novice at public engagement, I was learning on the job.

A colleague of ours visited most days to chat about how things were going in the negotiations. She had worked with the Hadley Centre before on the ecological impacts of climate change but she was here in Japan to represent the World Wide Fund for Nature which gave her observer status at the deliberations. Like a gossipy relative at an extended and fraught family gathering, she spent her days exchanging intelligence between the delegations, helping to oil the wheels of their discussions by giving the key players a sense of what the others were thinking. Apparently, progress was being made. But it was clear there were a lot of details to be sorted out in a very short time.

An important principle at stake behind a potential Kyoto Protocol had been laid down in the UN Framework Convention on Climate Change signed by the world's governments in Rio five years before. This was that countries from the developed world, whose emissions to date were by far the largest contributor to current climate change, would have to take the first step. So far though, a voluntary commitment for these countries to stabilize their greenhouse gas emissions at 1990 levels had failed to work. And what's more, there was a growing realization that climate change would not be solved unless emissions were not just levelled off but cut back drastically over the course of the twenty-first century. Here in Kyoto, over two short weeks, an agreement on how to do so would have to be thrashed out.[3]

One evening, Cath and I decided to sample the atmosphere of the negotiations. We closed down the stand for the night and sneaked into the main hall, taking two of the back row of seats reserved for the IPCC delegation. In front of us, under a high vaulted ceiling, stood regimented rows of chairs and tables facing a long table on a raised platform from which Argentinian ambassador Raúl Estrada-Oyuela chaired the meeting. Nations were arranged in alphabetical order from the front of the room. The delegations were provided with a handful of seats each and a placard of the country's name which the chief negotiator could place in the vertical position in order to indicate they wished to make an intervention.

Being exposed to the politics of climate change was a sobering experience. In the couple of hours we were there, government representatives spent their time discussing a single sentence of the draft Kyoto agreement. After the excitement of the rest of our day explaining our rapidly developing scientific understanding to eager delegates, this was a depressing reminder of how bogged down these crucial negotiations had become. We had a copy of the current text in front of us. Most of the words were contained in brackets, contentious phrases that were still the subject of dispute. The deliberations remained mired in the minutiae of a complex set of issues. The mood in the air chimed with what we were hearing from our insider from the World Wide Fund for Nature. The parties were still a long way from agreement.

The concepts of Emissions Trading and Joint Implementation lay at the heart of the difficulties.[4] While it was agreed that countries would have to reduce their emissions, it was also accepted that different nations could have different targets depending on their individual circumstances. For some, reducing emissions would be easier than others, if they were already phasing out coal-fired power stations like in the UK with its 'dash for gas' in the wake of privatization of the electricity industry, or planting new forests and in the process drawing down carbon dioxide from the atmosphere as was happening in Sweden, Finland and New Zealand. But some countries, the United States among them, wanted even more leeway in meeting their emissions targets.

Emissions Trading would mean that a country could team up with another developed country that was emitting less than their Kyoto targets and buy its right to emit under the Protocol. The economies of the former Soviet Union had crashed following the fall of the Berlin Wall in 1989. Here was an opportunity for some nations to duck responsibility for reducing their own emissions by subsidizing Russia and her former satellite states in rebuilding their economies using cleaner fuels.

Joint Implementation would mean that a country would get credit for helping a developing country keep down its emissions by helping it wean its energy generation off dirty technologies. Just like under Emissions Trading, this would allow a country like the United States to avoid some of the emissions reductions signed up to in Kyoto. As long as it helped other countries reduce their emissions, the proponents of Joint Implementation argued, it would be reducing global emissions overall and therefore would be allowed to do less to reduce its own polluting activities.

Much of the developing world was bitterly opposed to such arrangements. Climate change had been caused by the industrialized world, they argued, and it was up to them to fix it. Developing nations wouldn't have to make emissions reductions themselves. But the idea that developed nations could wriggle out of cutting their own emissions rankled. The rich north had benefited all these years and now it was they who should fix the problem directly, not profit from giving sweeteners to the poor south or exploit the happenstance of economic collapse in Eastern Europe. Emissions Trading, Joint Implementation, divisions between north and south; these were the buzzwords during those fractious, frantic days in Kyoto.

As if arguments between nations who accepted the science weren't difficult enough, some people were doing their best to make nations doubt the scientific basis for the negotiations. One of the best known was an American called Fred Singer. I had heard he was here in Kyoto, touring the delegations and trying to undermine their resolve. I suspected that sooner or later he would pay a visit to the only people at the conference centre presenting the scientific evidence on climate. I awaited his arrival at the Hadley Centre stand with trepidation.

Fred Singer was an American physicist who had founded a lobby group called the Science and Environmental Policy Project, which was affiliated with the Washington Institute for Values in Public Policy that was itself financed by the Reverend Sun Myung Moon's Unification Church.[5] Moon was a controversial figure whose church held mass wedding ceremonies, who was a passionate anti-communist and who invested huge sums in the *Washington Times*, a newspaper that published many articles, including by Singer, rejecting the scientific consensus on climate change, ozone depletion and the harmful effects of passive smoking.[6] Singer's lobby group propounded the view that environmental threats were manufactured by environmentalists who, he thought, had a 'hidden political agenda' against 'business, the free market and the capitalist system'.[7] To fight what he saw as excessive regulation of the polluting activities of industry, he argued that science was corrupt and scientists could not be trusted. To support his aims, Singer's group worked with other climate deniers like Patrick J. Michaels to try to discredit well-established scientific understanding. With strong financial resources and powerful political backers, such lobby groups had already succeeded in distorting public understanding of the climate problem.

Singer and colleagues – including Frederick Seitz who had founded the George C. Marshall Institute, another group promoting climate change denial – had singled out one climate scientist in particular for attack.[8] In letters published in leading American newspapers, such as the *Wall Street Journal*, they accused Ben Santer of making unauthorized changes to his chapter of the IPCC report after it had been accepted by governments. As a result, they claimed, Ben had deceived policymakers into believing the evidence for human-induced climate change was stronger than it really was. According to Singer, Ben had undertaken a 'crusade to provide a scientific cover for political action' by 'misusing the work of respected scientists'.[9]

As well as an attack on Ben, this was also an assault on the reputation of the IPCC, the likes of which it had not seen before. The UN body responsible for advising government on climate change was being accused of corruption. If the scientific arguments could seem abstruse to many people, a charge of corruption was one that many people could easily relate to. Except, it was a charge entirely without foundation.

The latest IPCC report, with its groundbreaking conclusion that there was a 'discernible human influence on global climate', had been produced exactly in accordance with the agreed procedures. The summary part of the report had been agreed word by word by governments and it was entirely consistent with the finalized version of Ben's chapter. The baseless accusation that the conclusions had been distorted were easy to rebut. But attempts by the scientific community to do so were rebuffed.

A response to the *Wall Street Journal* by Sir John Houghton, the leader of the entire report, was edited on publication to make it appear that Ben alone was responsible for the IPCC finding of discernible human influence on climate.[10] In fact, this had been a consensus decision supported by thousands of scientists and hundreds of governments. And when forty of Ben's colleagues joined him in also penning a response, their signatures were removed to make out it was only Ben responding to the allegations of distortion.

Singer followed up with a further attack that was even more shocking.[11] He accused Ben of 'scientific cleansing'. 'Cleansing' was a word in use at the time to refer to atrocities in Bosnia. It was a resonance not lost on Ben whose grandparents were subject to ethnic cleansing as victims of a Second World War concentration camp.[12] He sent an email to colleagues expressing his outrage that his actions were being equated with the act of genocide and stating that he found the accusation of 'scientific cleansing' to be morally repugnant.[13] His view that Singer's accusations were odious in the extreme was widely shared by the email's recipients.

The cleansing charge referred to a decision by Ben and the IPCC author team not to show temperature changes over just a few years at a time.[14] Instead, they included an illustration of temperature changes over several decades, during which time both climate models and observations showed comparable rates of warming. An equivalent illustration of temperature changes over just a few years, during which time temperatures can warm or cool purely by chance, would have been irrelevant. Only over longer periods do you expect to see the effects of climate change. It was the same point I had made in my talk at the American Geophysical Union in San Francisco. Climate change is a long-term process. 'No matter how loudly Dr S. Fred Singer broadcasts his visions of political tampering and scientific cleansing', Ben wrote in his email to colleagues, 'he cannot halt the inexorable progress of the science itself.'[15]

I felt confident that science would progress eventually, whatever the deniers did. But Ben's experience showed how disruptive such people could be and how unpleasant it could be to try to face them down. I knew what Fred Singer looked like from pictures. But I wasn't sure I knew what to expect if he came to call. All I could do, as Cath and I worked long hours at our Hadley Centre stand, was carry on talking with more reasonable people and wait for any approach.

It took a few days but eventually I saw him coming towards me as I stood by our pile of brochures waiting for the next enquiry. He was a squat man with twinkling eyes who greeted me with an enthusiastic handshake. It was me, the expert in attributing climate change, he wanted to talk to, rather than Cath, who specialized in projections of future climate. What have you got here, he asked me? It was a simple opening, an invitation to explain what I had to offer, little different from many of the other opening remarks I'd dealt with since starting work in Japan. In response, I launched into my standard account along the lines I had given hundreds of time before.

I explained the evidence we had that humankind was already changing climate. I showed him our projections of climate change in the future. And I described the increasing hunger, greater water scarcity and rising sea levels expected under modest-sounding increases in global temperature. He listened politely, but with an indulgent expression as if I were telling him fairy stories created to scare the children.

Our PC was showing an animation of how our climate model simulated temperatures of seawater. He seemed more interested in this. Global warming results from an enhanced greenhouse effect which means that additional heat is trapped in the climate system by rising atmospheric concentrations of greenhouse gases. But because water has a much greater capacity to soak up heat than air, much of the extra heat goes into warming the oceans rather than warming the atmosphere. Slowly the ocean warms from the surface, with extra heat penetrating deeper and deeper as the decades roll by. This is what you expect from basic physics and this is exactly what we saw when we watched our movie of ocean heating.

It was also what measurements of water temperatures taken from beneath the surface of the oceans were beginning to show.[16] Warming spreading to deeper and deeper levels was a telltale signature of global warming. It was still early days for monitoring changes in marine climate, but that particular fingerprint was starting to appear.

Singer didn't buy it. How could I rule out the effect going the other way, he asked, of the warming at the surface being caused by transfer of heat from below?[17] That would have nothing to do with greenhouse gases trapping additional heat in the atmosphere. Singer's explanation was ridiculous. If correct, it would mean the oceans would be cooling and there was no evidence of that. It was even more ridiculous when there was a simpler and far more likely explanation – for which there *was* evidence – already available. But, unfortunately, I couldn't yet prove that my explanation was correct.

There had been ocean-going measurement campaigns but they were still few and far between. Ships had sailed across the Pacific sending down thermometers on wires deep into the watery darkness. But such ad hoc efforts didn't provide a global picture like our climate models did, nor did they provide a set of regular snapshots in time of how the warming was gradually reaching further and further down below the surface.

A future targeted measurement campaign from ocean-going floats would resolve the question once and for all by showing irrefutable evidence that the world's oceans were warming in the manner expected. But Singer was exploiting current scientific uncertainties to further his cause of climate change denial. According to him, there was no reason to act on climate change because the evidence was insufficient. I couldn't convince him otherwise. Singer had drawn me on to his ground, one on which I didn't have all the answers.

As he left, I felt deflated that I hadn't won the argument. Scientifically based evidence, I knew, was the only sensible basis for climate action. And we had brought plenty of evidence with us to Kyoto. But Singer was peddling pseudoscience, half-baked theories with no shred of evidence to support them. I had met someone who wasn't prepared to change his mind in the face of new facts. That wasn't the mindset of a sceptical scientist. Yet with his confident manner, many delegates *could* see him as a reputable scientist, spreading plausible-seeming doubt on the reasons for tackling climate change. That was depressing, especially when I knew that Singer had form on another global environmental problem.[18] It was the ozone hole.

This dramatic loss of stratospheric ozone above the South Pole, first discovered in 1985 by Joe Farman and colleagues from the British Antarctic Survey, came as a big surprise to scientists and one with worrying implications given that stratospheric ozone protected life at the surface from damaging ultraviolet radiation.[19] Data from NASA satellites showing a similar catastrophic loss had previously been dismissed as being implausibly extreme.[20] Instead, thanks to low-tech monitoring from the ground and data from weather balloons, Farman convinced the scientific community that this was indeed a real effect. It just lacked an explanation.

Susan Solomon, an atmospheric chemist at the National Oceanic and Atmospheric Administration, provided one. Based on her understanding of the process of catalysis, whereby a chemical reaction is accelerated by the presence of a substance that remains untouched by the reaction and so can act repeatedly, she hypothesized that industrial chemicals called chlorofluorocarbons – CFCs for short – were part of just such a reaction. By reacting with ice clouds high up in the cold polar stratosphere, she proposed that CFCs were releasing chlorine in a form that could act as a catalyst capable of rapidly destroying ozone under the action of ultraviolet light during the polar spring. Not content with her groundbreaking theoretical work, Solomon then embarked on a quest to prove her ideas experimentally by leading expeditions to McMurdo Sound in Antarctica in 1986 and 1987. There, her team measured extremely high levels of chlorine oxide, the catalyst that was destroying ozone. She had found the proof that CFCs were a dangerous pollutant causing havoc with the natural world.[21]

CFCs didn't just threaten the ozone layer above Antarctica where few people lived. These damaging chemicals were also depleting stratospheric ozone at other latitudes, putting the health of billions at risk from skin cancer. This was a planetary threat that required coordinated global action. In Montreal, Canada in 1987, a protocol was signed that committed countries to phasing out their use of CFCs. Thanks to scientific consensus and government action, CFCs in spray-can propellants, fridges and air conditioners were replaced by roll-on deodorants and more energy-efficient refrigerants. Gradually, the stratospheric ozone layer started to recover, paving the way for the eventual repair of the Antarctic ozone hole. The Montreal Protocol demonstrated that political action, based on scientific understanding, could solve a global environmental problem.

Before he was a climate sceptic, Singer was an ozone hole sceptic. Together with Michaels, Ben Santer's opponent in San Francisco, he propounded a theory that ozone depletion was nothing to do with CFCs but instead was a natural consequence of volcanic eruptions.

Michaels and Singer made some calculations that appeared to show that volcanoes released far more ozone-depleting chemicals into the atmosphere than CFCs. According to their analysis, the eruption of Mount St Augustine in 1976 released more than 500 times the amount of ozone-damaging chlorine than the chlorine produced that year by the entire global production of CFCs. They publicized their results in a series of articles in the *Washington Times*, a newspaper founded by the Reverend Moon's church who funded Singer's lobby group. They also promoted their theory in US Congressional hearings.[22]

However, Singer and Michaels had made a simple error of interpretation. Atmospheric measurements clearly showed that the vast majority of chlorine emitted by volcanoes never made it as far as the stratosphere where the ozone layer sat, because the chlorine-containing ash was rained out before it even got there. CFCs, on the other hand, were chemically inert and able to travel up to the stratosphere in large quantities.[23]

Volcanic chlorine was irrelevant as far as ozone depletion was concerned. Their theory was easily refuted. It was scientific hokum. But to sympathetic lawmakers in Congress and gullible journalists, it sounded plausible enough and gained sufficient traction to hold up efforts to phase out CFCs. To help delay regulation, lobbyists for the chemical industries accused the scientists linking CFC pollution to the ozone hole of being corrupted by self-interest and anti-free-market ideologies.

Despite the efforts of Singer, Michaels and the other ozone hole sceptics, regulators soon banned CFCs and industry found less harmful replacements. Thanks to these actions, three decades later stratospheric ozone above Antarctica is now starting to increase. While it could take another forty years for the ozone layer to heal completely, the prompt elimination of the pollution responsible has minimized an important threat to life on earth.[24] But although they had failed to delay action on CFCs for long, Singer and his cohort of science denialists were having more success with global warming.

At the climate negotiations in Kyoto, I had seen for myself how Singer was exploiting scientific uncertainties to justify inaction on climate when I had lost a debate about ocean temperatures and what they meant for global warming. But if he had bested me here, I should remember Singer had been proved demonstrably wrong about the cause of the ozone hole. Despite the disappointment of my encounter, I hoped one day to have another chance to debate the scientific case on climate change with such a figure. I just didn't know when.

Nor did I know whether there would be a successful outcome to the political negotiations taking place a few yards away from where I had met one of the world's most prominent climate deniers. The haggling over the commitments of each of the developed countries occupied much of the negotiators' time through the days and nights that followed. It was painfully slow but they were making some progress. It would have been relatively plain sailing were it not for the contentious topic of whether developed nations could get out of their commitments through paying others to take up the slack. It was this that came within a whisker of scuppering an agreement.[25]

As the talks dragged on into the early hours of the morning following the final day scheduled for the talks, and with the cleaners waiting impatiently in the wings to ready the centre for its next congress, the developing nations were still refusing to accept Emissions Trading as part of the Kyoto agreement. Abortive efforts were made in the middle of the night to contact the Indian prime minister in the hope he could broker a deal. Instead, in a moment of high tension, as exhausted delegates realized they had finally reached the crunch make-or-break moments of the summit meeting, chairman Estrada-Oyuela deleted whole reams of existing text, the result of hours of haggling. Instead, he replaced it with three short sentences stating that trading of emissions would be allowed but must not replace domestic action. He read the new sentences out twice, very slowly, and then paused to wait for any reaction.

As Estrada-Oyuela banged down his gavel to formally enter his compromise formula into the Protocol, India raised their flag to object. The chair of the meeting ignored it and carried on to the rest of the document. He knew, as did everyone else there, that any country that openly challenged his authority would be responsible for single-handedly bringing down the negotiations. Everybody else kept quiet. To widespread relief, so did India. They had backed down. But although developing nations had lost this particular round, they were still gunning for Joint Implementation.

Developing countries hated Joint Implementation. To them it meant that developed countries could weasel their way out of their Kyoto commitment in return for token investments in the poorer nations, investment that would deliver financial benefits not to them but back to the rich donor nations. Developing countries *were* keen, however, on the idea of financial penalties being levied as a punishment for Kyoto targets being exceeded. These penalties would have to be used to develop clean technologies under what was named the Clean Development Fund. Realizing this Clean Development Fund had the same overall effect as Joint Implementation, the Americans spotted an opportunity. In what became known as the Kyoto 'surprise', the Americans and the developing countries found themselves in agreement, providing the Clean Development Fund became a mechanism and this new mechanism was there to contribute to compliance rather than punish non-compliance. Joint Implementation was ditched and replaced by the Clean Development Mechanism. Finally, chairman Estrada-Oyuela approved the final words of the newly born Kyoto Protocol.

An agreement running to twenty articles, many of which contain numerous clauses, it was a relatively complex document at first sight. But at its heart lay a set of groundbreaking commitments to reduce the greenhouse gas emissions of the developed nations. Now sufficient countries would have to endorse it through their internal political processes before the Protocol would enter into legal force. Both America and Russia were set to face stiff resistance in their domestic legislatures. But that was an issue for another day. Japan had hosted the most momentous international meeting ever held on climate change, one that would influence the lives of millions of people around the world for decades to come.

It had been fascinating being there when it happened, if only as a junior researcher trying to do my bit to keep delegates informed about the scientific evidence. It had been stressful at times, knowing I had to simplify the science while staying faithful to the facts. I'd been given a glimpse into the world of political horse-trading involved in coming to a major international agreement. I'd seen how uncertainties in scientific understanding and our determination to reduce them could be turned against us. And, though I didn't know it, I hadn't seen the half of it yet.

3

Strengthening the science

During the year following my trip to Kyoto, I spent a lot of time arguing. I listened to the positions of others. I set out my own views. Then we tried to settle our differences. I wasn't arguing with climate deniers but with my close colleagues.

We didn't yet have a complete explanation for the evolution of global temperatures over the last century. The scientific community needed one fast, and we at the Hadley Centre wanted to be the first to find it. This was the reason for our passionate debates. We wanted to write a scientific paper together.[1] But there was so much more at stake than that.

Thanks to the Kyoto agreement, the world had taken its first tentative step towards, one day perhaps, halting climate change. The targets to reduce emissions were nowhere near enough: they applied only to developed countries and only over the next decade. Even then it was not certain that countries would stick to them, minimal as they were. But at least Kyoto had shown an international willingness to engage seriously with the climate issue. Now, having made this initial move, countries would have to up their ambition in reducing emissions, or else face a future level of warming that could still be catastrophic.

To support their efforts, governments urgently needed more scientific advice about the consequences of those emissions. For that we needed to figure out how much of the warming so far was due to human activities, rather than natural variations. The fingerprints of human influence had been detected in atmospheric data. But that didn't tell us how quickly climate was changing, and therefore how substantial emissions reductions needed to be in future to avoid the many devastating consequences that most countries were desperate to avoid.

To figure it all out, the team of Simon Tett, Myles Allen, John Mitchell and I had been joined by William Ingram, a mathematician by training who had been tasked with including solar effects in our climate model. For hours at a time, we sat in John's office criticizing each other's positions while searching for robust conclusions. It seemed like no matter how hard we tried, we couldn't agree on what we were going to say.

We were using a novel technique which, for the first time, allowed us to calculate the contributions of natural and human-induced factors to observed warming. It was the technique that Myles had first described at the American Geophysical Union conference in San Francisco.[2] If it made our calculation possible, it wasn't making it easy.

Over the last hundred years, global warming hadn't followed the gradual rise of greenhouse gas concentrations in the air. Instead, warming was particularly rapid during the first half of the twentieth century before levelling off during the 1950s and '60s and taking off again in the 1970s. That early twentieth-century warming was a puzzle, because in those earlier years of industrial activities when emissions were much lower than they are now, greenhouse gas concentrations in the atmosphere were increasing more slowly.

Simon and I carried out thousands of years' worth of simulations of the Hadley Centre's climate model to try to understand the changes in atmospheric temperatures since global records began more than a hundred years ago. In separate simulations of the climate model, we included the main factors that could have affected atmospheric temperatures. Greenhouse gas concentrations increased in the model, warming its climate. Air pollution and the resultant aerosols cooled temperatures especially during the middle part of the twentieth century. Explosive volcanic eruptions ejected dust high into the stratosphere causing occasional cooling bursts. And solar variations, William's particular interest, had their own role to play.

An eleven-year cycle in the sun's output was discovered in 1843 by German astronomer Samuel Heinrich Schwabe, by analysing the waxing and waning of dark spots on the solar surface.[3] New analysis of sunspots showed that these cycles weren't completely regular. Some appeared to be slightly more energetic than others. It seemed that during the first half of the twentieth century solar output gradually increased. In the grand scheme of things, the sun's output has been remarkably steady for millions of years, so much so that the irradiance from the sun has been named the solar constant. But nevertheless, given the powerful nature of the sun's heating on terrestrial climate, even tiny changes could conceivably alter global temperatures. William included that effect in the climate model as well.

Now we could apply our new method for attributing climate change to its causes. We treated observed changes as a combination of multiple causal factors – greenhouse gas increases, changes in air pollution, solar variability and volcanic eruptions – each of which we had included in our various simulations of the climate model. Up until then, previous methods could only tell whether one particular factor, such as an increase in greenhouse gas concentrations or solar output, was or was not significant. Now we could work out how much it contributed in combination with all the others, whether it played a substantial role or was just a minor player. If the model responded too strongly or too weakly to a particular influence, we could also take that into account. If, for example, we saw a pattern of temperatures varying in concert with the eleven-year solar cycle in both models and observations, but there was a more vigorous variation in observations, we could scale up the model's vapid response. And finally, we could check for whether our climate model had an adequate representation of natural climate variability from processes like El Niño, in which surface waters in the Eastern Pacific become much warmer than normal for months and years at a time.[4]

We had thought that all of this invention would lead to a much clearer picture of the causes of climate change. Instead, we spent hours debating in John's office, while colleagues in surrounding offices overheard the raised voices and wondered what was going on behind the closed door. It was the warming in the first half of the twentieth century that was the root of all the trouble. Our results allowed two possible explanations. According to our analysis, the early-century warming could be mainly due to increasing solar output. Or it could be largely due to a combination of greenhouse gas warming and a multidecadal oscillation in ocean temperatures related to variations in the Gulf Stream. No matter how hard we looked for a decisive reason to prefer one explanation over the other, we couldn't find it.

To us, this wasn't good enough. Rather than have two competing explanations, we wanted to keep searching for the key piece of evidence we'd missed that would elevate one over the other. But in our attempts to look more and more closely at the finer details of simulated and observed temperature changes, we only stumbled further and further into the mire. If Simon found a comparison that favoured one explanation over the other and Myles agreed with him, William raised an objection that the test was incomplete. If I suggested another possible test, Simon saw problems with it and Myles proposed yet another method of discrimination. And on it went, while our boss John Mitchell acted as referee, pulling us up on any illogicalities and trying to keep the peace. When we reached stalemate, we agreed to meet another time to have another go.

Eventually we came to the only possible conclusion, the one that our passion had blinded us to all along. In the latter half of the twentieth century, we had clear results pointing to human activities as the main contributor to global warming. Then solar variability did not increase sufficiently and the ever-increasing concentrations of greenhouse gases in the atmosphere had a dominant effect. But in the early half of the twentieth century, we had two possible explanations – one that pointed to a greater role for solar changes and one that pointed to a greater role for El Niño and other natural climate oscillations. With our current climate model, it simply wasn't possible to tell them apart. At last, it dawned on us. Faced with scientific uncertainty we couldn't currently reduce, we should stop arguing and submit our results to an academic journal. We would propose both possibilities.[5]

The journal we chose, *Nature*, was a long shot. In 1998, the year Simon submitted our paper, there were 7,820 other submissions of which only 10 per cent finished up being published.[6] But we were in luck. *Nature*'s editors were convinced of our paper's originality, scientific significance and widespread relevance. They sent it out to a trio of anonymous reviewers who asked for improvements but thought its contents worthy of widespread dissemination. A revised draft of the paper, 'Causes of twentieth-century temperature change near the earth's surface' by Simon Tett, Peter Stott, Myles Allen, William Ingram and John Mitchell, was accepted for publication.

We were thrilled, Simon most of all, given his leadership of the writing. For me, it was a long-held ambition to be an author of a *Nature* paper. More importantly, we thought we had provided a convincing explanation for the temperature changes that mattered most, those over the past five decades when human-induced climate change had really taken hold. This was a big moment for the advancement of climate science. But there was a still bigger prize that remained elusive. Even though our main conclusion, that human factors explained much of the observed warming over the last fifty years, was straightforward, our findings could be dismissed because we hadn't accounted fully for the warming in the first half of the twentieth century. If we were going to convince sceptical peers, not to mention a hostile public, we needed evidence that was more definitive.

Worldwide, governments were still wrestling with the implications of Kyoto, trying to get their parliaments to ratify their commitments and persuade their citizens of the importance of man-made climate change. With huge financial implications at stake, they needed to be convinced that there was a robust scientific basis before radically altering the means of energy generation, the basis on which modern economic prosperity had been built.

Nature paper or not, we hadn't yet come up with the goods. What we needed was a new climate model, a model that could more realistically simulate weather systems and ocean currents, that could better simulate climatic variations. Such a model might provide the more compelling account of global warming we still lacked.

A new climate model would need a bigger computer. When I joined the Hadley Centre in 1996, we had access to a huge amount of computing power. But still, we didn't have enough. Fair weather cumulus on a summer's day, towering cumulonimbus in a stormy sky, the Gulf Stream trailing warm and salty water in a narrow stream along the North American coast: all are phenomena that span just a few kilometres across. Our computer wasn't powerful enough for our model to represent individual clouds or ocean eddies. Instead, it had to simulate the effects of many clouds, or lots of eddies, the effects of which it averaged over large areas, more than 250 kilometres across.[7]

By averaging over such large areas, we paid a price. For thousands of years before the Industrial Revolution, the concentrations of atmospheric greenhouse gases remained constant and there was a balance between incoming energy from the sun and outgoing energy radiated back to space. Global surface temperatures hardly changed. When we simulated this era in our climate model, with greenhouse gas concentrations held constant, we expected modelled temperatures to stay steady too. But instead, they changed.

The model's temperatures warmed up or cooled down because the exchanges of heat between the ocean and the atmosphere were incorrect. It was a problem that could only be fixed by forcing the flows – or fluxes – of heat between atmosphere and ocean in the model so that they stayed close to their observed values. At the time, this approach, known as flux correction, was used by all the world's climate modelling centres.[8] It worked, after a fashion, but nobody thought it was a totally satisfactory solution.

Keeping our model close to current reality was the only way to generate believable projections of climate change. But ramping up greenhouse gas emissions promised a very different reality, one of melting ice caps and vanishing forests. A climate model with flux correction wouldn't provide realistic predictions of the future, if that future turned out to be vastly different from today's. And given today's reality was already altered from that of a hundred years ago, it wouldn't necessarily do a good job of simulating past climate change either. If we wanted to stop using flux correction, we needed a bigger computer and we needed to rethink how our climate model worked.

To visualize how a climate model works, I like to imagine sitting on the ground at the bottom of an imaginary box of air, stretching off to north and south, east and west of me, with similar-sized boxes arranged in all directions and others piled in a stack above my head. The temperatures everywhere in this virtual box of mine are exactly the same. Other boxes upwind, downwind and skywards have different weather. Boxes in the upper troposphere have lower temperatures than at the surface, while each box in the lower stratosphere is warmer than the one beneath thanks to the warming power of stratospheric ozone. I can imagine a storm brewing, out to the west. The neighbouring box is already windier. A few boxes further westward, temperatures have dropped substantially and it is raining heavily.

Now imagine you know, right now, the weather in all those boxes all around the world. Thanks to the basic laws of physics you have the means to calculate the weather in the future. Atmospheric mass must be conserved. As must momentum. And temperatures must rise and fall in response to heating and cooling. The primitive equations – the unromantically named trinity of formulations that express balance in mass, momentum and thermal energy – predict the position of the storm fifteen minutes hence. With a new starting point for a second set of calculations, a forecast for fifteen minutes from now becomes a forecast half an hour ahead, a forecast for half an hour ahead becomes a forecast for forty-five minutes' time, and so on. Time-stepping its way forward our model can project days, years and even decades ahead. It is a discretized approach, partitioning space into chunky blocks and time into finite steps.

This approach to predicting the future was invented by Lewis Fry Richardson during the First World War.[9] L.F. Richardson was a mathematician with a passion for seeking relationships in the world of nature as well as the world of humankind.

He initiated the scientific analysis of conflict by analysing the relationship between the borders of countries and the probability of their going to war. In his researches, he tried to measure the coastlines of countries and discovered that the measured length of a coastline keeps getting longer and longer the more you try to take account of smaller and smaller wiggles in coastal indentations. It was the start of the science of fractals, objects which have structures that repeat as you look at them in greater and greater detail. And L.F. Richardson also began the science of weather forecasting.

Richardson was a committed pacifist who refused to enlist during the Great War and instead found himself ferrying the injured and dying back from the front as part of the Quaker ambulance unit.[10] Despite his long, harrowing and dangerous days in the theatre of conflict, he still had the energy to spend his resting time at night making a weather forecast by means of laborious arithmetical calculations. Just like modern forecasts, he computed the weather at a future time based on atmospheric conditions at the current time and knowledge of the primitive equations of meteorology. It took him several years to predict one day into the future. One day, he imagined thousands of human computers making the calculations, all working in harmony, passing their latest results back and forth from grid box to grid box across the surface of the earth and up through the depth of the atmosphere. Given the number of calculations needed and the speed with which they needed to be made before any forecast was superseded by events on the ground, numerical weather forecasting was not yet a practical proposition.

It only became so with the advent of numerical rather than human computers in the 1950s. And with the resulting rapid advances in numerical weather forecasting, came the first steps in numerical climate prediction. Those early advances were made at the Geophysical Fluid Dynamics Laboratory in Princeton, New Jersey, which had access to the most powerful computers of the time.[11]

Researchers including Norman Phillips, Syukuro Manabe and Richard Wetherald were the pioneers, producing the first computer simulation of the global atmosphere's circulation in 1956, the first credible simulation of the earth's atmospheric climate in 1967 and the first climate simulation of both ocean and atmosphere in 1975.[12] In the 1980s, those efforts were joined by the National Center for Atmospheric Research in Colorado and a section of NASA located in New York City called the Goddard Institute for Space Studies (GISS) where James Hansen, the scientist who called out global warming to Congress in 1988, was its director.

It was Hansen's team at NASA GISS who demonstrated for the first time that accurate prediction was possible on climate as well as weather timescales, over a matter of years not just days.[13] Working fast following the explosive eruption of the Mount Pinatubo volcano in 1991, the researchers made a forecast of how climate would evolve over the next decade. Their prediction was that global temperatures would dip for a couple of years before recovering as a result of dust thrown high into the stratosphere by the eruption reflecting sunlight back to space and cooling the planet until the dust gradually settled back down to earth. The prediction proved to be remarkably accurate and represented a major breakthrough for climate science. By demonstrating that climate models could accurately anticipate the climatic effects of volcanoes, their success also showed that numerical models could potentially forecast the climatic effects of increasing greenhouse gas concentrations, changing solar output and industrial air pollution.

By the late 1990s, things had moved on a lot since Richardson's days. But while forecasting the weather over the next few days was proving increasingly successful, simulating climate change over the next century meant using much larger grid boxes than in the weather forecasting model. At the Hadley Centre, if we used grid boxes smaller than 250 kilometres across, too much computer power would be spent making the calculations needed for each time step. Our climate simulations, like the weather simulations Lewis Fry Richardson made by hand all those years ago, would simply take too long to make.

It was thanks to a colleague called Geoff Jenkins that we were able to speed up our calculations. Geoff wasn't a research scientist but he understood what we were doing and he talked to the people who held the purse strings. A regular visitor to gatherings of our research group, Geoff was a bearded and inquisitive figure who sported colourful shirts and clashing ties. His job was to represent our work to our government funders. To help make his case, he made sure he was always up to date with what we climate modellers were up to.

When Geoff put forward a case to Whitehall that a new and bigger supercomputer would help deliver the improved scientific understanding the governments so desperately needed, the paymasters listened. Thanks to the resultant increase in computer power, and some ingenuity, we were able to eliminate the need for flux correction in our newest climate model.[14] It completely revolutionized our approach to predicting climate change. The ingenuity gave us new ways to represent the motions of clouds and mixing of seawater. As a result, atmospheric winds and oceanic eddies in the models were much better represented. At the same time, the increase in computing power allowed us to decrease the size of the model's ocean grid boxes by a factor of six. This meant that ocean currents like the Gulf Stream were also much more realistic.

Our new model was called HadCM3, short for Hadley Centre Coupled Model Three.[15] The model's own fluxes of heat between ocean and atmosphere now agreed with reality and no longer needed to be corrected. It simulated a stable climate in the era before cars and coal-fired power stations when greenhouse gas concentrations were held constant. It gave us hope that we could solve the puzzle of what caused the warming in the early part of the twentieth century and establish a more robust explanation for the climate change seen over the last 140 years.

In return for helping us obtain a new supercomputer, Geoff set us a grand challenge. It was a challenge we resisted at first because it was not the normal way of doing things. And it felt like a gamble we didn't need to make. Geoff suggested that we should include all the most important factors that could affect climate in a single set of simulations of our brand-new model, HadCM3. His idea was that we should set our model going in 1860 and see whether it could reproduce the observed evolution of global temperatures: the stagnation during the late nineteenth century, the warming in the early twentieth century, the second period of relatively stable temperatures in the 1950s and '60s and the current surge in warming culminating in a new global temperature record in 1998. Generally speaking, temperatures warmed faster over the land than over the ocean. And following the eruption of the Mount Pinatubo volcano in the Philippines on 15 June 1991, global temperatures dipped sharply for a couple of years before recovering. What Geoff wanted to know was: could HadCM3 predict all that? Such a model would map the story of climate change in the modern era.

This would be a demanding test of our new model's performance. It would also be very expensive, because of chaos. Chaos is the process by which tiny perturbations in the initial state of a dynamical system, such as the oceans and atmosphere, explode into big differences later on.[16] A butterfly fluttering its way along the Gambia River in West Africa turns left rather than right and several days later a Category 5 hurricane swings towards Florida rather than dissipating harmlessly over the Atlantic Ocean. You can't expect to predict the weather precisely decades from now because you can't specify today's conditions with the infinite precision required, butterflies and all. But averaged weather is climate. And that you can hope to forecast over future decades, if you take account of the possible permutations in temperature, wind and rain that chaos could produce. To take account of chaos, we would have to set four runs of our climate model going, each from slightly different initial conditions, each set free to chart a different course through the myriad possible weather patterns consistent with the natural and human climatic factors included in the model. Comparing an ensemble of simulations with the observations, we would be able to tell if our model was capable of predicting climate.

Surprisingly, perhaps, Simon wasn't keen. By considering all the relevant climatic factors in the model separately, he thought, we had all we needed to attribute their relative contributions to global warming. We could set the model running with only natural factors, including from volcanic eruptions and changes in solar heating, and we could compare those runs with climate model runs which included only the human factors on climate. Simon didn't think we needed to put the model through a beauty contest by seeing how well it did when it included all the factors together. But I found Geoff's argument persuasive. I could see that a wider audience would be interested in whether a model could simulate all of past climate change, whatever caused it, whether natural or human.

I hoped that this approach would make our assessments of climate change more accessible to a wider range of people. It was a simple question to ask: could our model reproduce the history of global temperatures? In answering that question, I imagined generating a striking illustration showing why climate had changed, one that could potentially capture the imagination of policymakers and members of the public. I didn't then realize quite how striking that illustration would turn out to be.

It was agreed I would make the runs. They were going to take several months to complete. I would need to monitor progress at evenings and weekends and restart the model if, as it was prone to doing, the supercomputer crashed. And day by day, as my runs inched their way through the years of the nineteenth and twentieth centuries, I would have to monitor how my model was doing, how my simulated temperatures compared with the real ones. This was exciting; a virtual-reality ride through 140 years of changing climate. I couldn't wait to get started.

It was also a virtual adventure with real risks. At this time, in mid 2000, policymakers were eagerly awaiting the next report of the IPCC due to be published early the following year. They needed it to come to clear, robust conclusions about the causes of climate change. If HadCM3 succeeded in simulating the complexities of past temperature changes, it would be a clear demonstration of the power of humanity to change its environment. If it failed it would be a big setback, not just to the Hadley Centre's mission to develop a reliable climate model, but to international efforts to deal with the challenges of climate change.

With a keen sense of anticipation, I set up the runs of HadCM3. We named them the 'all bells and whistles' runs because they factored in all the most important effects on climate. They included the increasing concentrations of greenhouse gases. They incorporated the climatic effects of sulphate aerosols, which had made the atmosphere hazier and clouds brighter, thereby reflecting more of the incoming solar radiation back to space. They took account of explosive volcanic eruptions which threw ash high into the stratosphere where it reflected incoming sunlight back to space, cooling climate for several years before the ash eventually fell back to earth. And they included the subtle heating and cooling effects from small variations in solar output.

Our groundbreaking model also generated a wealth of natural complexity all by itself. Emerging every few years from the primitive equations that expressed in mathematical form the fundamental physical realities of earth's climate, El Niño conditions would appear in the model, bringing warmer than usual surface waters to the Eastern Pacific and slackening the trade winds along the equator. On a slower timescale, the surface waters of the model's Atlantic Ocean would warm and cool for decades at a time as the Gulf Stream sped up or slowed down. Such oscillations were also possible explanations for some of the temperature changes seen over the last century.

Each new run of the model was automatically allocated a five-letter moniker by the software that scheduled the simulations. By a freaky coincidence that felt like fate playing its hand, my all bells and whistles runs were given the initial three letters A, B and W. I launched the run named ABWEA with atmospheric and oceanic conditions from a date in 1860. Shortly afterwards, I fired the starter pistol on its three siblings, whose initial atmospheric and oceanic states were taken from different days in the same year and which the scheduler had called ABWEB, ABWEC and ABWED.

Predicting climate over 140 years was going to take the best part of three months. If the model did badly we would still have to report the results. Such a failure would be manna from heaven for the climate deniers, an unexpected opportunity for them to bash climate predictions based on climate models' apparent inability to simulate the past. And our government funders wouldn't be best pleased. They'd invested a lot of money on our model and our computer, money they'd want spent well. There was a lot riding on this. But all I could do was wait and see what happened. Success or failure lay in the lap of our very expensive supercomputer.

Arriving at my office each morning, the first thing I did was find out how my model runs were doing. With another twenty-four hours elapsed, my fast-forward through time had spooled on another two years. Each day I could add a few more data points to the graph comparing simulated temperatures with reality. Between 1860 and 1900, global temperatures varied from year to year but didn't warm or cool overall. This generally stable climate was well captured by the ABW ensemble. So far, it seemed, so good. This was reassuring but not the acid test. It was the sustained changes over the twentieth century that were going to be the hardest to simulate.

The substantial warming between the turn of the century and the 1940s was a puzzle that John, Simon, Myles, William and I had been unable to resolve using the previous climate model. Now HadCM3 was showing increasing global temperatures after 1900 just like observed. It looked promising. I began to think we might have an explanation for the enigmatic warming during the first fifty years of the twentieth century.

The years ticked by in my all bells and whistles simulations until I could assess the next stage of their global warming examination. During the 1960s and '70s my runs showed a prolonged pause in warming, just like reality. There were still warmer years and colder ones as El Niño and its sister La Niña blew hot and cold across the vast expanses of the model's Pacific Ocean. But observed temperatures didn't warm overall, and neither did the model's.

The supercomputer was being kept busy by our determination to understand the causes of past warming. As well as my ABW runs, John had authorized a parallel set of runs in which only natural factors on climate had been included. These simulations failed to capture warming in recent decades. Now HadCM3 faced its biggest test yet. Could the model simulate the correct rate of warming from the late 1970s when it included human factors?

As I waited for my runs to complete, I had time to think about how best to visualize the results. Up until then, the general rule in John's team had been to keep our graphs separate. If we wanted to look at how HadCM3 simulated climate when it included only natural factors, the obvious thing to do was to plot temperatures from the model and compare with the observed temperatures. If we wanted to compare observations with how the model simulated climate when it included human factors, that would be a different plot. As the ABW cohort crept towards the finishing line, I had a simple yet novel insight. I decided to plot temperatures from the different types of simulations on the same piece of paper.

I showed the final result to John. At the top of my A4 sheet I had plotted a graph showing how temperature varied from the start of 1860 to the end of 1999 in observations and in model simulations including just natural factors. The sustained warming seen since 1970 was not captured by the model. In the middle of the piece of paper I had plotted the same observed temperature evolution, but this time compared to model simulations with only human factors. These model simulations failed to capture the early-century warming. At the bottom of the page, I plotted the comparison between the same observations and the all bells and whistles simulations. Now that they had finished their simulation of twentieth-century temperature changes, I could at last assess whether the model had captured all the temperature changes since the mid nineteenth century. It had. It was hard to tell observations and models apart.[17]

Seeing the delight on John's face as he looked at the plot brought home to me what a stunning achievement this was. Over the last few months I had concentrated day by day on the tasks needed to make the simulations. But now that I could reflect on what I'd done, I felt the same mixture of relief and pride I'd experienced when I had passed the viva examination for my PhD. That feeling was already quite a long way in the past. Without a doubt, *this* moment was the best of my scientific career so far. But to capitalize on the opportunity that had been presented to me, there was no time to waste. I would need to write up my results for a scientific journal.

I needed to describe how the HadCM3 experiments provided a comprehensive explanation for the variations in global temperatures seen at the surface of the earth. The warming in the early twentieth century was due to several factors acting in concert. It was a time of the earliest stirrings of greenhouse warming, of no major volcanic eruptions to cool the planet, and of an increasingly active sun gradually raising solar radiation arriving at earth. The mid century hiatus in warming was attributable to the global effects of air pollution, thanks to the coal-fired smogs from the industrial and domestic chimneys of the Western world which shielded the atmosphere from incoming solar radiation, and temporarily held back global warming. Finally, starting in the late 1970s with the introduction of legislation to clean up air pollution, the warming effects of greenhouse gases won out over the cooling effects of aerosols. The greenhouse gases being pumped into the atmosphere were steadily warming the climate. Without reductions in emissions, this was a warming that would become more and more difficult for humanity to cope with.

My illustration, the three sets of lines on a single sheet of paper, provided the clearest demonstration yet seen that recent warming was man-made not natural.[18] It was vital, not just that this conclusion was published, but that it appeared in the next report of the Intergovernmental Panel on Climate Change, due to appear early the following year. This was the most significant international scientific assessment yet made by the IPCC, the first since the Kyoto meeting of 1997 and a crucial source of information for policymakers seeking to build on the Protocol signed in Japan.

For this new report, John Mitchell had taken on the role played by Ben Santer the previous time in leading the writing of the chapter on the causes of past climate change. Unlike Ben, John faced an additional challenge. This time, the Summary for Policymakers, the part of the report which summarized the main findings in a short document, would include a small number of illustrative graphs as well as text. John needed an illustration that summed up the current understanding of the causes of climate change. My plot fitted the bill perfectly. But this only redoubled the importance of my work getting through peer review quickly and then out to the public. Only if graphs were supported by published papers could they appear in this crucial report.

Submitting my work to the journal *Science* was a risky option. The American equivalent of *Nature*, this was regarded just as highly internationally and rejected just as many papers. John knew that if my paper didn't make it through peer review by the end of the year, he would have to remove this vital piece of evidence from the IPCC's summary document. But he thought I deserved a crack at my own first-author paper in a leading journal. He was prepared to take the risk.

I submitted my paper on 11 September 2000. In recognition of his encouragement to carry out the all bells and whistles runs, I included Geoff among an author list that also took in Simon, Myles, John, and the newest member of our attribution group at the Hadley Centre, Gareth Jones, all of whom helped interpret the results and write up the conclusions. The next few weeks turned into an anxious wait. Just like when I was monitoring my all bells and whistles runs, there was nothing I could do to influence the outcome. This time success or failure lay in the hands of two anonymous expert reviewers.

The letter from *Science* arrived in my in tray just before teatime. I walked into the corridor outside my office where colleagues were congregating for their afternoon break, clutching my missive and sporting a massive grin. My paper had been accepted for publication subject to some easily dealt with revisions.[19] John looked mightily relieved. His gamble had paid off. He had his illustration to take to the crucial IPCC meeting in Shanghai in January 2001.

The Shanghai meeting was the final stage in a three-year process of compiling the latest assessment. It was here that the scientific authors of the report would have to agree every word of the Summary for Policymakers with representatives of governments from across the world. Without a successful conclusion of this meeting, there could be no report. And the encounter turned out to be as nerve-shredding for John Mitchell as the equivalent gathering had been for Ben Santer five years earlier in Madrid.[20]

Agreeing the illustration showing the causes of climate change was straightforward. Nobody could object to a graph that was fully peer-reviewed and published in a leading journal. Redrawn for the Summary for Policymakers, it now depicted the observational time series as a wobbly thick red line and the ranges of modelled temperatures by grey bands. Only when the model included human factors did the red line representing the observations fit within the grey band representing the model. It had a dramatic impact and a simple interpretation: human-induced climate change was behind the current warming trend.

It should have been easy to agree a headline sentence to go with it. But words are more easily quoted than pictures. Government delegates knew the IPCC's statement on the causes of climate change was the most eagerly awaited and would be the most widely cited of the whole report. The sentence was potentially much more contentious than the illustration.

John Mitchell's original proposal stated 'it is likely that increasing concentrations of anthropogenic greenhouse gases have contributed substantially to the observed warming over the last 50 years.'[21] A basic consideration of the physics of the greenhouse effect implied a substantial warming would result from the rapidly increasing concentrations of carbon dioxide and other greenhouse gases in the atmosphere. HadCM3 clearly showed greenhouse gases dominating the warming in recent years and this was backed up by a wealth of supporting evidence collected by fellow members of the International Detection and Attribution Group. John's statement should have been uncontroversial. Indeed, most governments found it so. But when the sentence was put before delegates for agreement, one government objected.

It was Saudi Arabia who refused to support it, on the grounds that 'substantial' did not translate into Arabic.[22] The summary documents would be translated into the six official working languages of the United Nations, so it was reasonable for representatives who spoke French, Spanish, Chinese, Russian and Arabic to keep an eye on how the phrases agreed in English would appear in translation. But objecting to a particular formulation could also be a useful negotiating tactic.

Sir John Houghton, leader of the report and chair of the meeting, had no alternative but to call for a 'contact group', a meeting of interested parties to convene in a separate room to thrash out a form of words everyone could agree to. Closeted in a side room, the Saudis asked John if the scientists could state how much of the observed warming was due to greenhouse gases increasing. This put John and his fellow chapter authors in a tricky position. Substantial was a word chosen to indicate that the contribution from greenhouse gases was rather large but without stating exactly how much. My *Science* paper clearly showed that greenhouse gases had a major impact. But as to exactly how much, that was harder to say, especially since there was little scientific literature that precisely quantified the greenhouse gas contribution.

In the end, John was saved by use of the word 'likely' in the sentence, a word that in IPCC circles came with a precise definition. 'Likely' meant that there was a greater than 66 per cent chance of the statement being true. All the scientists had to do, John realized, was to find a fraction of the observed warming that they were confident could be attributed to greenhouse gas increases with a greater than 66 per cent chance of being correct. 'Most of the warming' seemed to cover it. To some of the scientists, this sounded stronger than their original formulation of a 'substantial warming'. But the peer-reviewed evidence they had clearly showed that the chance that greenhouse gases caused less than half the warming was rather small.

The new formulation was acceptable to Saudi Arabia and back in the main room the revised headline statement was approved. 'In the light of new evidence and taking into account the remaining uncertainties,' it stated, 'most of the observed warming over the last 50 years is likely to have been due to the increase in greenhouse gas concentrations.'

Later that night, as Houghton watched the Chinese New Year fireworks from his hotel balcony, he reflected on a meeting that had been relatively straightforward compared with the bun fight in Madrid.[23] But once again the attribution statement was the most contentious, and once again the attribution statement attracted the most attention in the international media the next day. For governments engaged in the UN process to avoid dangerous climate change, the scientific evidence that mattered more than anything else was the extent to which climate change could be blamed on human activities.

A few days later, John held court during the tea break in our corridor at the Hadley Centre, telling us about his time in Shanghai. He was eager to expand on his culinary adventures, as well as tell us about the pressure-cooker environment of an IPCC approval meeting. Duck's feet I could do without but I found the process of agreeing a scientific assessment in such a pressure-cooker atmosphere fascinating. While turning scientific findings into text that could be agreed in consensus by over a hundred governments sounded hard work, I could see the advantage in such an approach. Having agreed what the science said, now countries had to get on dealing with climate change without getting sidetracked by rhetorical debates about the evidence.

The IPCC's Third Assessment Report marked a major landmark.[24] A much stronger link between human activities and climate change had been established than was possible five years before when the 'discernible human influence' statement had been agreed. Scientific progress had been considerable. Flux correction had been eliminated. Climate models could simulate the past evolution of atmospheric temperatures. Greenhouse gas increases were indeed the dominant factor controlling global climate change. All this evidence, so expertly martialled by the authors of the newest IPCC report, made me feel much more optimistic that international cooperation would lead to increasingly serious efforts to tackle climate change.

Encouraged by our scientifically robust findings, many countries *were* making plans to reduce their greenhouse gas emissions under the Kyoto Protocol. But just as the science had evolved in the years since Ben Santer was attacked by the oil-funded lobby groups, so too had the politics. On the last day of the IPCC meeting in Shanghai, a new American president was inaugurated in Washington. George W. Bush had beaten Al Gore in a wafer-thin election result that, in a political version of chaos theory, could have gone the other way were it not for the happenstance of poorly designed butterfly ballots in the crucial swing state of Florida.[25] As a result, the White House was now occupied by a man who viewed climate change very differently from his vanquished Democrat opponent.

The Russian presidency had changed too. Vladimir Putin was now in charge, a ruthless operator who wouldn't hesitate to leverage Russia's diplomatic advantage by holding out on ratifying the Kyoto agreement. The Kyoto Protocol would only come into legal force once fifty-five countries had endorsed their agreed emissions reductions through their national legislatures, and when those countries accounted for more than 55 per cent of the total carbon dioxide emissions by developed countries in 1990.[26]

Kyoto was going to depend on Russia. In due course, British researchers would be called to the heart of Moscow to face a politically inspired inquisition into the alleged falsehoods of our science. Like Ben Santer before me, I was going to find out what it's like to speak truth to power, when that power does not want you to tell what you know.

4

Ambushed by power

My first steps on Russian soil were like arrivals at other airports. There was a queue at immigration while a uniformed official inspected my passport, and a wait by the baggage carousel wondering whether my bag would eventually appear. It felt reassuringly familiar. But as I stepped outside, that was about to change.

Earlier that morning – 6 July 2004 – I had arrived at Heathrow Airport to catch the flight to Moscow. I was part of a group of British scientists being flown out to discuss climate science with our counterparts in Russia. I was looking forward to the trip, my first opportunity to take part in a diplomatic rendezvous to help build closer scientific ties between two nations. I had important research to present and I was looking forward to hearing about the latest findings from Russian colleagues at the country's foremost scientific organization, the Russian Academy of Sciences. We would be welcomed by their president, Professor Yuri Osipov, and, for the first time, I would meet the UK's most influential scientist, the government's Chief Scientific Adviser Sir David King, who would head our delegation.

But then I saw Geoff Jenkins at check-in and discovered there was dramatic news. He greeted me with a remark about cloak-and-dagger tactics. The long-agreed agenda for our meeting had been ripped up, he told me. Instead, according to the information Geoff had received late the night before, a roll call of many of the world's most prominent proponents of climate change denial had been summoned to Moscow to confront us.

It was a shock. I'd seen the emails between the Foreign Office, the Russian Academy and ourselves setting up the arrangements; deciding on the topics for discussion, drawing up a list of speakers and confirming hotels and restaurants. Our trip was supposed to be a bilateral discussion about our latest climate research. Now, judging by the new list of attendees Geoff showed me while we queued to collect our boarding passes and deposit our bags, we were facing a multinational inquisition into the case for human-induced climate change.

The roster included Richard Lindzen, an American professor from the Massachusetts Institute of Technology who expounded far-fetched contrarian theories about climate change. Piers Corbyn, the eccentric promoter of a crank weather-forecasting procedure who attacked the Met Office's peer-reviewed weather and climate science, would also be there. They would be joined by William Kininmonth, a retired meteorologist from Australia who rejected the scientific consensus on climate change, Nils-Axel Mörner, a Swedish academic who believed that sea levels were not rising in contradiction of all the evidence, and many others, coming from far and wide, all of them known for their bizarre rejection of the mainstream view that greenhouse gas emissions were warming climate. Rather than the Russian Academy president being responsible for the meeting, it now seemed it was another Academy member in charge, somebody called Yuri Izrael.

I couldn't help wondering whether this last-minute change of plan might be an elaborate hoax. Was it really credible that this man, Yuri Izrael, presumably behind Yuri Osipov's back, had arranged for such well-known climate deniers to be flown in from around the world, just for a two-day meeting? Geoff suggested we'd better try to prepare a response, just in case. In the departure lounge I rang the office to see if somebody could provide information about Corbyn's theories while Geoff phoned Cath Senior to ask her to send us something on Lindzen's claims.

At the departure gate, we were joined by Sir John Houghton, who had led the previous Intergovernmental Panel on Climate Change (IPCC) report on the science of climate change, and Peter Cox, a Hadley Centre colleague who had modelled the effects of greenhouse gas emissions on the biosphere. With little time for more than a collective expression of surprise at the new circumstances we found ourselves in, we were boarding our plane to Russia. Allocated separate seats scattered throughout a full cabin, we had several hours alone to mull over what might happen next. It gave me time to think about what we could say to rebut the denier arguments that could be put to us.

Fortunately, I had just co-written a paper reviewing the latest advances in understanding the causes of climate change with my colleagues from the International Detection and Attribution Group.[1] Phil Jones had joined the group, a shy English academic from the University of East Anglia, and had brought new weather records and measures of past climate – showing just how unusual recent warming was. Overall, our findings showed that the strength of evidence for a human influence on climate had grown since the last IPCC report. I had brought a copy along and I tried to reread it, to make sure I was well versed in the relevant information. But I couldn't concentrate.

I looked around at the mainly Russian passengers around me. The man at my side was watching a film on a portable DVD player, a novel luxury at the time. A woman walking down the aisle had the words 'sex, love, money' brightly emblazoned across her T-shirt. Naïve though it might have been to think so, it looked like some of my fellow travellers wouldn't have felt out of place in Putin's kleptocracy of oil-rich oligarchs. Who had disrupted this meeting of ours, I wondered? What were they trying to achieve? And how would I fare, if I were pitched into a battle to defend my science against attacks from experienced and effective climate deniers?

On leaving the arrivals hall in Moscow airport, we were ushered over to a people carrier by a local employee of the British Council. His instructions were to drive us to our hotel but John Houghton wanted to go straight to the British embassy so we could discuss our plan of action. As we swept down a dual carriageway into the city, we asked our driver to change course. But he couldn't – or wouldn't – understand. We wanted to contact the embassy for a Russian speaker who could explain what we wanted to do. But we didn't remember the area codes we needed for our British mobiles. We were trapped, going in the wrong direction. It was alarming that just a few minutes into our arrival on Russian soil and over such a simple matter as being taken where we wanted to go, we had already lost control of events.

While Geoff and I frantically fiddled with our phones, John told us about Yuri Izrael, the new chair of the meeting. He knew him from working together on the IPCC. Apparently, he was a bully and a troublemaker. All the same, John thought, wasn't Izrael supposed to be in Geneva for a meeting of the World Meteorological Organization? Maybe this was all just a bluff, he said.

Eventually, somebody remembered the dialling code we needed and we were able to find a Russian speaker at the embassy who could redirect our driver. But she also gave us disturbing new information. The leader of our delegation, Sir David King, was on the verge of pulling out, and two other senior members of our group, Professor Nigel Arnell, an authority on climate impacts, and John Ashton, an ex-diplomat and policy expert, had been told not to travel. As we approached the city centre, grey buildings under grey skies, we discussed the implications. If there was going to be a show trial of climate science, as we now suspected, and the British government wanted nothing to do with it, why hadn't we too been stood down?

The embassy, a glass-fronted modernist building by the river, had concrete roadblocks outside and a police car stationed by the entrance.

We passed through a gatehouse where our bags were scanned and entered the main entrance where the senior diplomat in charge of our visit escorted us past a lecture room where the Foreign Secretary, Jack Straw, was giving a speech. Our host led us up a flight of stairs and showed us into her spacious office. She invited us to take a seat and sat down behind her desk ready to talk.

Copies of the new agenda were distributed around the group. The original programme that had been agreed weeks before consisted of a one-day high-profile segment of leading Russian and British climate scientists, followed by a lower-key second day for younger scientists. This new version lasted a day and a half and featured talks from all the international climate deniers we'd heard about earlier, interspersed with contributions from the official British delegation. Most significantly, the programme featured Andrei Illarionov, Putin's senior economic adviser, who was slated to make some opening remarks.

According to the diplomat, the new agenda was no bluff. But the government's Chief Scientific Adviser had still not decided whether to come. She was grateful we were there, and since we were, the next day's meeting might as well carry on. Sir John wasn't convinced. If Sir David didn't arrive, and with others staying at home, that would leave a depleted team in the firing line. His concerns were listened to politely. But the UK government's position was made quite clear. We were here to defend the science.

Put simply, the Kyoto Protocol would only come into legal force once Russia had ratified the agreement in its national parliament. Illarionov was pushing hard for this not to happen. He believed Kyoto would cripple his country's economy and had recently described the Protocol as an 'economic Auschwitz for Russia'.[2] To support his case, he was promoting the views of climate deniers, namely that climate change was a natural phenomenon.

Now that Illarionov had overthrown the Academy's original intentions for the crucial meeting, it was down to us, the depleted group of British scientists that had made it there, to put forward the counter-case: that past warming really was man-made, that only reducing emissions as envisaged in the Kyoto Protocol could halt the dangerous impacts – floods, droughts, storms and sea level rise – that would threaten humanity's collective well-being and security. From our perspective, the stakes could not have been higher. 'We work for the Queen,' was Geoff's position. 'If required we will carry on.'

With time running out for the Chief Scientific Adviser to catch the last flight out, the diplomat was on the phone to London trying to establish his movements. Meanwhile, we looked at the briefings we'd requested earlier and which had now been emailed over and printed off. They were damning about both men's ideas. Corbyn's claim to be able to predict the weather weeks in advance by analysing solar activity was based on a crackpot theory that had never been published in a peer-reviewed journal.[3] Lindzen's hypothesis that the climate would heal itself by reducing the amount of high-level tropical cirrus clouds, thereby allowing more heat to escape to space, contradicted all the evidence that pointed to changes in clouds enhancing, not reducing, greenhouse warming.[4] Our briefings dealt with two of the specious arguments that could be put to us the next day. Thinking about the many more that Lindzen, Corbyn and company could throw at us, I felt frustrated that nonsense like this was being given such high-level credence and angry that it could derail the Kyoto agreement. But all I could do was continue to prepare material for our rebuttals as we waited anxiously for news of our head of delegation.

At last the call came through: Sir David would be flying out. Apparently, he had held discussions with Academy president Yuri Osipov agreeing alterations to the meeting agenda to bring it more into line with what had been originally intended. It was good news. Now, with our head of delegation confirmed and some moves towards a compromise on the meeting, we could head over to our hotel to regroup over dinner.

We reached the hotel restaurant after a journey squashed together in a small private car, one of the unofficial network of so-called gypsy cabs that had been flagged down for us by Russian-speaking embassy staff. The trip involved several U-turns and frequent doubling back along the city's wide boulevards, and by the time we arrived we were more than ready for something to eat. Discussion over dinner turned to politics. Listening to our British embassy colleague, it sounded like a new Cold War was looming. The last independent television channel had recently fallen under government control. Embassy staff were tailed, even on holiday. Rooms were bugged.

In bed that night, it took me a long time to fall asleep. I felt ill prepared for what lay ahead. Having to defend the well-established findings of all of climate science felt like a daunting prospect. If Illarionov and the climate deniers were involved, the careful accumulation of scientific data, peer-reviewed scrutiny and independent lines of evidence could count for nothing. I could defend my work in the court of science. But I didn't see how we were going to defend ourselves if this particular court was going to get political.

At 8 a.m. the next morning, Sir David King arrived at our designated table in the hotel's dining room, looking every inch the influential mover in powerful circles. A private secretary hovered at his side, holding his briefing papers and wearing an air of fast-stream civil service efficiency. While we ate breakfast, our embassy contact described the letter she held in front of her. It was from Osipov and it provided confirmation that the two leaders would meet at 9 a.m. at the Russian Academy of Sciences. Their task would be to adjust the programme along the lines brokered by King the night before.

The Chief Scientific Adviser travelled to the Academy by car with his private secretary while the rest of us followed along in a chartered minibus. Crossing the Moskva River, we passed an overblown statue sited on a small island in the middle of the water. It depicted Peter the Great, right arm aloft and holding a golden scroll while standing astride a massive battleship. An ugly representation of militaristic might, it seemed to stand in opposition to what international agreements like the Kyoto Protocol were trying to achieve in working together peacefully across borders for the common good. The meeting we were travelling to could help threaten the Protocol ever entering into force and damage public understanding of climate change in Russia for years to come. Whatever happened over the next two days could be hugely important.

When we arrived at the Russian Academy of Sciences I saw that it too sported a massive sculpture, a futuristic metallic structure that adorned the roof of an oppressive-looking multi-storey block that towered over us as we walked towards the entrance. I wondered one last time whether this confrontation with the forces of climate change denial might not happen, whether our head of delegation would have made the Academicians see sense, whether we'd be back to something like the original bilateral meeting of scientists. Flying people here from as far afield as Australia and the US at such short notice for such a brief meeting seemed an extraordinarily extreme measure.

The wishful thinking didn't last long. We were greeted in the entrance foyer by a row of stern-faced women standing behind trestle tables. They dealt each of us a large brown paper parcel for us to carry through to the institute's main auditorium. Passing through the double doors I finally saw what awaited us. Lined up to the left behind a row of desks along a central aisle sat Lindzen, Corbyn, Kininmonth and all the others who figured on the agenda I had first seen the previous day. Ahead lay a central aisle leading to a raised platform and a long top table which currently stood empty. To the right sat parallel rows of desks to those on the left, also empty, waiting for our imminent occupation. We were in a debating chamber, like the House of Commons with government and opposition members facing each other across neutral territory.

We exchanged cursory acknowledgements across the divide and took up places on our side of the room at the desks which came equipped with microphones and headphones. On the far wall behind us interpreters sat within glass-fronted booths waiting to start simultaneous interpretation. Geoff Jenkins was nowhere to be seen.

With nothing else to keep me occupied I flicked through the material in my parcel. Under the brown paper covering lay a thick sheaf of papers including copies of all the presentations, a recent article in the *Daily Mail* dismissing climate change by botanist and bird lover David Bellamy, and a set of curriculum vitae of each of the participants. Each of the climate deniers was graced with a comprehensive set of awards, publications and committees. Ours were terse and written to promote suspicions of institutional bias. That of Michael Grubb, an energy expert who had joined us after a later flight from London, stated, 'Of course the UK government now pays him to say that we should reduce emissions, so he is unlikely to provide any independent views.'[5] Mine stated simply, 'Peter Stott is a climatologist at the Met Office's Hadley Centre for Climate Prediction and Research who has been frequently used by the UK and international media as a technical spokesperson for the Centre.'[6]

There was a tense silence. The top table remained unoccupied, our head of delegation still absent. I thought about possible responses to the deniers' points. I looked across the aisle and wondered what they were thinking about us. I waited for something to happen.

After about an hour, a British embassy employee came by each member of our delegation, one by one, looking grim and telling us to start packing our things. Five minutes later she was back, instructing us to walk out. The morning half gone, we found ourselves back where we started, in the foyer with the trestle tables and the stern-faced women. In front of us was a staircase leading down from the upper floors. David King walked down it towards us and asked us to gather round. I thought he was going to tell us we were leaving. Instead he told us he was leaving. We were going to stay.

Later Geoff told me what had happened after he found himself in a room with the leaders of the delegations. King was supposed to be agreeing a compromise with Academy president, Osipov. But instead, the encounter turned into a shouting match between King and Illarionov. Each refused to change their proposed agenda. Each accused the other of bad faith.

In a trice, we were back in the main auditorium, once again in our positions to the right of the aisle, our paper parcels once more on the desks next to our headphones. But this time we were listening to our leader make a short statement that he was going to leave to meet the Russian science minister. Then he strode out accompanied by his private secretary. Now it was up to us.

Yuri Izrael, a big bear of a man, was in position at the top table next to Andrei Illarionov, who sat stony-faced to his right. On our side of the room, Geoff sat to my right and to the right of him John Houghton took the place nearest to the head of the room, our positions reflecting our respective status in the hierarchy of seniority. On the opposite side facing us sat all the climate deniers waiting for battle to be joined.

We kicked off by listening to the translation of Izrael's opening remarks through our headphones. The Russian Academy had invited a number of distinguished international experts, he told us, but the British government's Chief Scientific Adviser had disrupted and delayed the meeting by refusing to accept their participation. The meeting would now proceed as detailed in the programme.

At this point, Michael Grubb tried to intervene. He could not be present the following day as he had a sick relative to attend to at home. Would it be possible for him to make his presentation today? What he had to say was very relevant to the debate about the impact of Kyoto on the Russian economy. Izrael refused. It was exasperating and unfair but there was nothing we could do other than protest, which our chairman took as another example of the disruptive tactics of the British. He had lain down a marker for the sort of biased chairing we could expect from now on.

John Houghton had been given the first speaking slot. It was an opportunity for him to present the main findings of the most recent IPCC report. But despite having been emailed through in good time, his PowerPoint presentation was not available on the Academy's computer system. Instead he had to set out the evidence for man-made climate change without visual aids. This was alarming. Not only were we were being demonized, every effort was being made to prevent us from getting a fair hearing.

While John was talking, Peter Cox booted up his laptop, copied the speaker's presentation to a memory stick and handed it to one of the Academy's technicians. At last, displayed on the screen at the front of the room, we could see the presentation. It was a small but cheering victory. As a result, just before the end of his slot, John was able to backtrack and show two key results.

First was the graph that I had produced for the Summary for Policymakers of the IPCC report.[7] It showed how observed warming was *not* consistent with climate model simulations that included only natural factors but *was* consistent with climate model simulations that included increasing greenhouse gas concentrations. It was a clear demonstration that recent climate change was mainly due to human activities.

Second was another illustration from the Summary for Policymakers. This showed how temperatures averaged over the northern hemisphere had changed over the last millennium and had been reconstructed from tree rings, ice cores, corals and other proxy records of climate from the pre-instrumental era and thermometer readings when such measures became widely available from the mid nineteenth century onwards.[8] The graph had a distinctive shape, with variations of temperatures from year to year but without any systematic warming until the early twentieth century when temperatures started to shoot up. To North Americans familiar with the sport, it looked like the stick used by ice hockey players when it is laid on its side and the blade is pointing upwards. The then director of the Geophysical Fluid Dynamics Laboratory in Princeton, Jerry Mahlman, had named this iconic illustration of how global warming was an unprecedented modern phenomenon the hockey stick.[9] It was already one of the most controversial graphs in science.

Putin's right-hand man had been quiet so far but now he exploded into life. The hockey stick had been disproved, he insisted. Current temperatures were *not* warmer than they had been for 1,000 years, he maintained. To support his contention, he presented his own graph of past temperatures. This looked nothing like the one that John Houghton had shown us just a few minutes before. According to this alternative reconstruction, there was nothing unusual about current warmth. Instead, it had been surpassed previously, in the fifteenth century, during what was known as the medieval warm period.

This was just the start of a lengthy and angry intervention.[10] He did not have a speaking slot, not even in Izrael's revised agenda, but as if in response to his counterpart's perceived bad faith in delaying the start of the meeting and then leaving, it appeared that Illarionov was now engaging in some counter-disruption. His thoughts ran on to other matters. We were not seeing any high frequency of emergency situations or events, he declared, nor was there any increase in the number of floods or droughts. If there is an insignificant increase in the temperature, he told us, it is not due to human factors but to the natural factors related to the planet itself and solar activity. There was no evidence confirming a positive link between the level of carbon dioxide and temperature changes and if there was such a link, he claimed, it was of a reverse nature. In other words, he was saying that it is not carbon dioxide that influences the temperature on earth, but temperatures that drive carbon dioxide: temperature fluctuations are caused by solar activity influencing the concentration of carbon dioxide. To sum up, Illarionov was saying that none of the assertions made in the Kyoto Protocol and the scientific theory on which the Kyoto Protocol is based had been borne out by actual data.

While Illarionov was speaking, we suggested to the chair that it would be more efficient if we responded to each of his points as we went along. But Izrael would not permit it. Exasperated at yet another refusal to allow reasonable debate, we had no option but to keep a rein on our frustration and wait patiently for Illarionov to finish his lengthy address.

When we finally got a chance to say something, we wanted to start by countering his attack on the hockey stick. I knew about this, thanks to the paper I had brought with me on the plane, the one summarizing the latest understanding of the causes of climate change that I had co-written with Phil Jones and other colleagues from the International Detection and Attribution Group. John Houghton looked across at Geoff and Geoff looked across at me. They wanted *me* to take this one.

There was a lot I wanted to explain. The IPCC's hockey stick came from a reconstruction of temperatures published in 1999 by an American scientist called Michael Mann.[11] Mann's results showed that temperatures averaged across the northern hemisphere were now warmer than they had been at any time over the last 1,000 years. This important finding had been confirmed by further analyses by other researchers whose additional temperature reconstructions we had included in our paper reviewing the latest science. It was why I could be so confident that Illarionov was wrong. Subsequent research had shown that Mann's original conclusion was correct.[12]

Despite this, I knew why Illarionov had highlighted the hockey stick. Over the past year, it had come under sustained attack from climate deniers.[13] It meant I needed to explain why these attempts to rewrite the history of past climate, including the temperature reconstruction he had just shown us, were *not* correct. And I would have to refer to the work of an aerospace engineer called Willie Soon, and Sallie Baliunas, a Harvard astrophysicist, and two Canadians, a retired mining consultant called Stephen McIntyre and Ross McKitrick, an economist. Even though they had no expertise in climate, James Inhofe, chair of the United States Senate Committee on Environment and Public Works and recipient of generous funding from oil and gas interests, had promoted their results.[14] He invited them to testify before his committee and used their findings to help derail a crucial vote that would have seen America legislate to reduce its emissions of greenhouse gases. Their work had helped Inhofe's cause and now it was helping Illarionov's too.

Soon and Baliunas, with funding from the American Petroleum Institute, had carried out an analysis of previous published records of past climate which, they claimed, supported average northern hemisphere temperatures being warmer in medieval times.[15] But despite being published (in almost identical form) in two different journals, their analysis suffered from some scarcely credible flaws.[16] Soon and Baliunas interpreted records as supporting a warmer medieval period than present if they showed any single region as being unusually warm, wet or dry at any time over a period of many centuries. This was nonsensical: a region being unusually wet or dry didn't tell you anything about its warmth; you would need many regions to be warm simultaneously, not just one, for the whole hemisphere to be warmer than average; their criterion of 'unusual' was interpreted so laxly as to be totally meaningless. To make matters worse, many of the records didn't even cover recent decades. A comparison with recent warmth wasn't even possible.

Their analysis was so irredeemably flawed that the editor-in-chief and most of the editorial board of *Climate Research*, one of the two journals involved, resigned in protest at the actions of the publisher and the editor responsible.[17] The other journal, *Energy and Environment*, specialized in climate change denial pieces and had a rock-bottom scientific reputation as a result. Despite this, Inhofe invited Soon and Baliunas to testify before his Senate committee in July 2003 alongside Mann who, suitably outnumbered, did his best to defend his own work and the other well-established findings of climate science. And later that year came another politically inspired assault on the hockey stick.

Senators John McCain, a Republican, and Joseph Lieberman, a Democrat, proposed a bill requiring the Environmental Protection Agency to regulate American greenhouse gas emissions.[18] On 27 October 2003, three days before the crucial vote, *Energy and Environment* published another blockbuster paper.[19] This new effort was by the two Canadians, Stephen McIntyre and Ross McKitrick, and it too declared the hockey stick dead and buried. Only this time, the verdict, wrong though it was, was somewhat harder to rebut.

Rather than analysing previous work like Soon and Baliunas had done, McIntyre and McKitrick had produced their own reconstruction of past temperatures. They used some of the same proxy records as Mann had used, but their result was dramatically different. Instead of previous centuries being cooler in the northern hemisphere, they found that the fifteenth century was warmer. This new reconstruction, the same one that Illarionov had just shown at the Russian Academy of Sciences, looked pretty convincing at first. But after a bit of digging by Mann and others, it soon became apparent what they'd done. They'd ignored most of the proxy records relevant to the fifteenth century which invalidated their conclusion about earlier warming. And their critique of the original hockey stick, that it came from bad data, was down to them having misinterpreted a spreadsheet they'd been sent.[20] Conveniently for Inhofe, however, such mistakes had not yet been discovered when a fossil fuel-funded lobbying organization distributed a press release ahead of the Senate vote. 'Important global warming study audited – numerous errors found', it said; 'new research reveals the UN IPCC Hockey Stick theory of climate change is flawed'.[21]

The McCain Lieberman Bill lost narrowly, by 55 votes to 43.[22] It was impossible to know whether the defeat could be put down to the attacks on Mann's research and its prominence in the IPCC report. But the confected hockey stick controversy had surely played its part, just as it was playing its part here in Moscow. There was so much I wanted to contradict in Illarionov's lengthy diatribe. But I started by saying that McIntyre and McKitrick had made serious errors in their analysis.

From the top table, Izrael made to stop me even before I had finished my first sentence, gesticulating angrily in my direction. Appalled, I listened to the interpreters translating him saying that it was Houghton who must respond not me. From my right, Geoff, who was also clearly disturbed by this escalation in hostilities, urged me to carry on. I continued to say that other independent reconstructions of past temperatures all supported the contention that recent warmth was unusual in at least the last millennium. Now Izrael and I were talking over each other. I was not allowed to answer a question that was *not* directed at me, I heard through my headphones, Izrael's words rendered into English by the interpreter. Feeling affronted that I was not deemed worthy of addressing the meeting and angry that the question *was* directed at me as much as anyone else on our team, I determined to press on. According to a host of evidence, I told anyone who was still listening, the sustained nature of recent warming was not seen in any previous era. Still Izrael refused to give way, talking over me as I tried to describe some more of the relevant science. With all eyes on the top table where Illarionov sat watching on approvingly and with little point me carrying on, I ground to a halt. The debate about the hockey stick had descended into a carefully confected farce.

We had to settle the issue of who should and who should not be allowed to speak. I sat back and listened to the debate about whether I would be allowed to present my explanations. Geoff Jenkins, supported by John Houghton, argued vigorously in my defence. Izrael, backed up by Illarionov, spoke out against. Being argued over like this was a novel experience. My status to address the meeting had become a point of principle to both sides.

Across the divide, the climate deniers watched the proceedings without comment. From our perspective, it was perfectly reasonable to ask the scientist present who knew most about the technical details to address the issue. But I wondered what they thought, and whether they were as angry as Izrael appeared to be at Houghton ducking the hockey stick question. Perhaps an argumentative and bad-tempered encounter suited them more. Probably both sides were being played if Andrei Illarionov's aim was to demonstrate that science had little useful to say about climate change either way.

In the end, Izrael allowed me to briefly repeat my two points, that there were well-documented problems with the Canadians' analysis and that since the IPCC report came out, a number of other independent reconstructions had been developed which agreed with the Mann result. It was getting late. Izrael wanted us to move on.

All this arguing had eaten up precious time. Our chairman still had two more speakers to get through before lunch, which was in any case going to be heavily delayed. The room was brightly lit at the front podium and along the aisle and the first two rows of seats, but the rows of seats behind were in deep shadow. Most of the members of the Russian Academy of Sciences, who had originally been slated to speak, had been marginalized, relegated to spectating from the back of the darkened room. But Georgy Golitsyn, director of the Obukhov Institute of Atmospheric Physics, emerged from the gloom to make a short presentation. Prefacing his remarks with a courageous statement that he had a different opinion from Illarionov and Izrael, he offered a sober and balanced presentation of the current state of climate science. We knew enough to act on climate, even if some scientific uncertainties still remained, he said.

For the morning session, like scheduling a headliner at a festival, Izrael had kept his star turn, Richard Lindzen, to last. Earlier in his career, Lindzen had commanded widespread respect as an eminent scientist, tackling difficult questions in the atmospheric sciences and boasting an impressive track record of publications and prizes. Latterly he had moved far away from the mainstream and become a vocal advocate for the case that climate models were flawed and that headline statements from the IPCC were politically motivated and scientifically incorrect. His contrarian theories, including that clouds acted to cool climate under greenhouse warming, had been quickly discredited losing him credibility with the scientific community as a result.

Squat, with a grizzled beard and an ironic expression, Lindzen placed himself firmly on a higher plane from everyone else here today. Because nobody was labelling their graphs properly he maintained, he was not going to show any. Instead he had written out his argument in bullet points on a series of wordy slides.[23] Consensus was a dangerous, intimidating and ambiguous concept.[24] According to him, the IPCC conclusion in 2001 that 'most of the observed warming over the last 50 years is likely to have been due to the increase in greenhouse gas concentrations' was the most egregious example of such a dangerous consensus.

The IPCC conclusion, Lindzen argued, implied that as much as 4 degrees Celsius of global warming were being attributed to greenhouse gas emissions.[25] This huge level of warming, he claimed, was being balanced by a large 3 degrees Celsius cooling due to aerosols from air pollution, which gave the overall observed warming of 1 degree. The net result was that climate models predicted far too high rates of warming in future when air pollution was cleaned up and aerosols were eliminated. According to Lindzen, climate alarmism was based on a fundamental error, traceable back to the detection and attribution community.

I hadn't heard this argument before, but it sounded ridiculous, even if it came out of the lips of such an eminent and self-confident speaker. I wondered whether he had actually read any of our scientific papers. The HadCM3 climate model accurately simulated past temperature changes and the model didn't have anything like 3 degrees Celsius of cooling from aerosols. In fact, this cooling was only a few tenths of a degree. But the chairman did not allow us to respond, and Lindzen was able to press on to his second line of argument.

Surprisingly, this wasn't the climatic self-healing hypothesis through changes in clouds, perhaps because as Cath's briefing had shown, the weight of evidence was starting to stack up against it. Instead it was that warming should lead to reduced atmospheric humidity.[26] Because water vapour, like carbon dioxide, is a greenhouse gas, such reductions in humidity would reduce the warming resulting from greenhouse gas emissions. This seemed even more far-fetched than his cloud hypothesis and went contrary to long-established reasoning that atmospheric humidity would increase under warming as a result of increased evaporation from the oceans. This so-called positive feedback – positive because more emissions lead to more warming which leads to more water vapour and therefore even more warming – was a staple of atmospheric science textbooks. And yet here was the arch contrarian, Professor Richard Lindzen, declaring that the textbooks had got it all wrong, even in the face of overwhelming observational evidence to the contrary.

Finally came his summing up, delivered with the confident air of a leading barrister concluding his case before a jury. The Kyoto agreement would have no discernible impact on global warming, he claimed, regardless of what anyone believed about climate change.[27] Lindzen finished with a quote from Joseph Goebbels: 'If you repeat a lie often enough, people will believe it.'[28]

Quoting Goebbels, the Nazi minister of propaganda, against us felt like another deliberate ramping up of the provocations being thrown in our direction. We were accused of being corrupt liars. But with no time for questions and a very late lunch waiting, we were driven into the Academy canteen for a meal of thin soup, and meat and mashed potato. We kept our counsel away from our opponents sitting at another table and hoped we would be allowed to make our case when the session resumed.

After the break, Geoff Jenkins had an opportunity to state what was widely understood by the scientific community about climate change. The evidence was clear, he stated, that atmospheric temperatures were warming and that human activities were largely responsible. Further greenhouse gas emissions would lead to further warming. The result would be rising sea levels, melting sea ice and more extreme weather. There was also a risk that the uptake of carbon dioxide emissions by soils and vegetation would be seriously compromised. This would increase atmospheric carbon dioxide concentrations and accelerate climate change.[29]

Geoff's clear summation of the evidence was treated by Illarionov as a further provocation to which he responded with another stream of questions and challenges. Another lengthy argument ensued with Izrael as to whether we would be allowed to answer each point in turn. Small victories were won, as the chairman allowed us to address one or two specific points. But Putin's right-hand man was building up a case, brick by brick, that the scientific uncertainties were too great to justify countries reducing their greenhouse gas emissions. We were not being allowed to put an effective counter-case. It was immensely frustrating.

There was more obfuscation and censorship to endure before we were allowed to break for the evening. The original programme would have showcased the climate research of Russian academicians, who now sat disenfranchised in the shadows, silently watching the unfolding drama. Instead we were forced to listen to William Kininmonth, who had been flown all the way from Australia, to pontificate on why meteorology ruled out human-induced climate change, and Piers Corbyn, who presented a wild-eyed sales pitch for his outlandish and unverified solar-based weather forecasting technique.[30]

With the clock having long since ticked past the original finishing time, Michael Grubb tried to intervene, requesting an opportunity to make one or two points about the possible economic consequences of Kyoto for Russia, before he had to return home to his sick relative. Izrael refused, citing once again the disruption of the meeting by the British delegation. Yet again, an argument ensued about whether a British delegate would be allowed to speak.

Grubb broke the deadlock when he offered to delay his travel plans. Perhaps shamed into backing down, speakers from the other side of the floor offered to delay their presentations until the next day. Allowed to speak, Grubb adopted a deferential tone, expressing a desire that Russian economists carry out similar calculations to his, which showed that Kyoto would not harm Russia's economy as Andrei Illarionov feared. His conclusion was based on empirical evidence that greenhouse gas emissions do not tend to scale with GDP as economies expand. Reducing emissions would not mean that Russia would have to reduce its economic output because economies in development tended to adopt less carbon-intensive ways of generating energy.

As the working day came to its weary end, Illarionov finally appeared ready to engage in a genuine debate. A lively but relatively polite discussion ensued between the two economists about the links between greenhouse gas emissions and economic activity. As we left the hall it felt like the meeting had finished on a more positive note than it had started. Later, the mood would darken once again. But for now, I was pleased to be leaving such a claustrophobic environment.

It was half past seven in the evening and, despite the attractions of the vodka being offered at a reception in the Academy, we were whisked off to an alternative reception at the deputy British ambassador's residence with some of the Russian scientists who had been bumped off Izrael's programme. Drinks of Pimm's were offered from a large bowl at the front of a grand reception hall. The deputy ambassador's wife came over and apologized for the heavy rain which had put paid to her holding the reception in the garden, a rare attribute for a central Moscow house. It was mentioned that we should expect rooms to be bugged despite regular sweeps. Sir David King made a speech in which he referred to Maria Sharapova's recent Wimbledon win.[31] This young Russian winning the hearts of the British people made it a propitious time, he said, for building closer ties between our two people.

Resuming the next morning at the Russian Academy, relations between the two groups were about to reach a new low. David King had been allowed to lead off the session with a presentation on the science of climate change. Not scheduled on anybody's agenda, this intervention took the best part of an hour. Unfortunately, in his presentation summing up the case for the seriousness of climate change, he had chosen to highlight the receding snows of Mount Kilimanjaro.

Now it was the climate deniers asking the questions, rather than Illarionov, and they seemed to know an awful lot about the weather records made by Brooke Bond Tea at their plantations on the foothills of the mountain. They had evidence that the position of the snowline on Kilimanjaro was not related to local temperature records. According to them, the explanation for the diminishing snow was a lot more complex than King was making out. I suspected they might be right.

Worldwide, it was clear that glaciers were receding. There could be no doubt about that, or that this global pattern was a consequence of global warming. Most glaciers were now melting more at their base than they were gaining from snowfall at higher elevations. But tropical glaciers were behaving differently. They were at such high altitudes that their snows were not melting but evaporating. It was the dryness of the air that mattered most for Kilimanjaro, not the summit temperatures which generally stayed well below freezing. And as to whether any increasing dryness in East Africa could be attributed to human-induced climate change, that question was still unanswered.[32]

In highlighting the receding snows of Kilimanjaro, King had chosen a weak example of climate change and the deniers knew it. Emboldened, they pressed him on other matters, on sea level rise and solar changes. The exchanges became heated. Eventually King lost his patience. 'I will not accept any more rude questions,' he told them. 'I have another engagement to go to and wish to be excused.' He began to leave. As he walked down the central aisle towards the exit, Illarionov angrily addressed his retreating back. 'No, you are not excused,' he shouted. 'It was you who disrupted this meeting, not me.'

Now it was the turn of two more members of the climate change denial team to speak and they took their opportunity to up the ante. Nils-Axel Mörner, emeritus professor from the University of Stockholm, delivered a passionate presentation detailing a mass of data claiming to show no evidence of sea level rise in the Maldives or globally that could be linked to greenhouse gas increases. Without the threat of flooding, according to Mörner, the IPCC tiger had lost its teeth.[33] Paul Reiter, an expert in insect-borne diseases from the University of Paris, claimed that malaria transmission had little to do with climate. He saw the IPCC as politically motivated scientific dogma, akin to the Soviet-supported dogma of Lysenkoism that rejected natural selection and saw thousands of mainstream biologists sent to prison, fired or executed under Stalin's reign of terror.[34] Reiter claimed that scientists who worked for the IPCC were like researchers who supported the use of eugenics as a means of discrimination in Nazi Germany.

We were being accused of being accessories to a conspiracy to pervert science and silence our opponents.[35] But without any sense of irony, Izrael refused to allow us to respond to the charges being laid at us from across the aisle. On our side of the room, we were getting restless, angry about the gross unfairness of the conduct of the debate but unsure what we should do in response. At my side, in contrast to my more agitated colleagues, Geoff was calmly writing in his notebook, preparing the points he wished to make if he got the opportunity. At last, he was given one. Knowing that this could be our final chance to defend ourselves and aware that he could be cut off at any moment, he kept it brief. This was not the meeting to which we had been invited. The scientific evidence for climate change should be judged on its merits, not on distorted representations of the IPCC. Satellite data showed that sea levels were rising at over 3 millimetres per year. This was an expected consequence of the thermal expansion of seawater and the melting of snow and ice.

Mörner started banging the table with his fist, claiming he had been maliciously excluded from participating in the last IPCC report, a charge totally without foundation. Denied once again an opportunity to respond, and frustrated at being unable to defend ourselves against the accusations being hurled in our direction, both sides were now shouting across each other. In desperation, Peter Cox walked down the aisle, jabbing his finger in the chairman's direction, demanding he let us respond.

With Putin's right-hand man at Izrael's side, his chairing wasn't going to change. The most we could do until the meeting broke up for lunch was to shout across the aisle demanding that our adversaries justify their extreme opinions. When we finally broke up I felt too furious to risk being in the canteen alongside such people despite being chivvied along by a functionary of the Academy, a feeling shared by my colleagues who also refused to leave the chamber. But then we learnt that they had gone to attend a press conference to which our side wasn't invited. We could go and get something to eat after all.

The meeting had disintegrated. Illarionov had his chaos. We had witnessed the tactics of climate change denial writ large: the spreading of confusion and doubt about the science bolstered by intemperate attacks on the integrity of the scientists. Getting the truth a hearing was next to impossible in such an environment, despite the wealth of evidence we had to support our case.

While we ate, Illarionov presided over a press conference, with Izrael and the climate deniers. Britain was forcing governments to ratify the Kyoto Protocol against their will, the presidential adviser told the assembled journalists. 'To our great regret this is a war,' he told the reporters, 'This is a total war against our country; a war that uses all kinds of means.'[36]

For those of us left to participate in the shortened young scientists' segment, the afternoon saw peace break out. John Houghton, Geoff Jenkins and Peter Cox had left for the airport. With Illarionov, Izrael and the others elsewhere, the Academy was returned to its members. They could emerge from the shadows to present their findings in a spirit of mutual respect. They had the data to show that climate change was as much of a reality for their country as for elsewhere in the world. Russia's temperatures were rising fast, particularly in the Arctic regions. Permafrost was melting, threatening the stability of buildings in northern cities. Our Russian colleagues told us that they were appalled at what had taken place earlier, although only academician Georgy Golitsyn had felt able to show any dissent. Everyone had little doubt that greenhouse gas emissions were responsible for much of the global warming that had been seen.

They knew too that climate science was a fascinating subject with much left to discover and understand. Everybody was invited to give a talk, anybody could ask questions, nobody was dismissed for what they had to say. At the end of a relaxed afternoon, a photograph was taken of a smiling group of British and Russian researchers, assembled in the venue which only a few hours earlier had been the scene of so much rancour and ill will.

Later that evening, with colleagues from the UK who like me had stayed on for the young scientists' meeting, I went to a party in a refurbished warehouse. British ex-pats and fashionable Muscovites had gathered to mark the departure of the British Council's director in

Russia. His valedictory speech painted a picture of a country that, as I had experienced myself, was chaotic and threatening. The lift in which he had once been trapped for twenty minutes remained unreliable and unfixed. His offices had recently been raided by armed tax police.[37]

Later, over drinks and canapés, the outgoing director came over to have a word. He knew about our meeting and the difficulties with Illarionov. Clearly it had been touch and go whether the British delegation took part. Apparently, it was the prime minister who had settled the matter, insisting that the Chief Scientific Adviser travel to Moscow. Otherwise I might not have seen the climate deniers in action. I could have spent my time sightseeing instead.

I was finally able to see a few of the sights the following morning, before catching my flight home. On my own and free to choose my itinerary, I wanted to find the onion domes of St Basil's Cathedral and the forbidding walls of the Kremlin. I decided to take a stroll towards Red Square. As I walked through the streets, I thought about what I'd experienced these last few days. I had been part of a show trial, in which well-established scientific evidence had been traduced, attempts had been made to silence me, and the integrity of me and my colleagues had been attacked. I was deflated and angry. The experience had brought home to me that I was starting to become a prominent climate expert. But the arguments over whether I could address the hockey stick question showed me this knowledge wasn't always welcome. Having expertise could be dangerous.

It was my evidence, and that of my colleagues, that had convinced world governments to agree a treaty limiting greenhouse gas emissions with profound consequences for the planet. It was this agreement that one of Russia's most powerful men believed would cripple his country's economy, that according to him amounted to a declaration of war. And now, for the first time since I arrived in Moscow, I was alone, without the support of colleagues and British embassy staff. The sense of paranoia that had been mounting ever since I learnt that Illarionov had sabotaged our meeting, reached new heights. I had seen how badly such people wanted the scientific evidence on climate change quashed. Like lions chasing their prey, they knew that the best strategy was to target individuals who had been separated from the pack. That was their approach with Ben Santer. It could be their approach with me.

Nearing the city centre, I stopped to look at my map. A man wearing jeans and a leather jacket appeared beside me and pointed me the way to go. I couldn't be sure I was being followed, but the strange encounter made me keener to mingle with the tourist crowds in Red Square, buy a souvenir from the GUM department store, and return to my hotel where a British Council driver was waiting to take me to the airport. Arriving in the departure lounge, I still felt on edge. At last, walking down the jet bridge towards my plane, the sight of British Airways cabin staff about to inspect my boarding pass felt immensely reassuring.

As I approached the door, I noticed that a young Russian soldier in a green military uniform under a large peaked cap was standing next to the external staircase leading down to the tarmac. He looked at the passengers one by one as they entered the plane. As I stepped past him I saw him register my presence. Then, he turned and left. I was not the final passenger. But clearly, I was the final person of interest to the authorities. Sinking into my seat, I looked across at what my neighbour was reading. It was the centre-page spread of the *Daily Mail*, this morning's edition distributed free to inbound passengers. It showed the bearded and beaming face of botanist David Bellamy next to yet another article of his underneath a banner headline which stated 'Global Warming? What a load of poppycock.'[38]

If few people outside Russia would get to hear about the meeting at the Academy, returning home I was among a planeload of people being fed similarly biased climate misinformation. It was a vivid demonstration that the tentacles of climate change denial spread far and wide, facilitated by a supportive and popular media. With powerful vested interests working against us, it felt like the odds were stacked against climate science prompting change of the magnitude required, even to meet the Kyoto agreement, let alone the much more stringent

reductions in emissions that would be needed to avoid the worst effects
of global warming.

I was heading home wiser but sadder. The optimism that had taken
me to Heathrow three days ago had evaporated. Before the meeting
at the Russian Academy I had thought the threat of global warming
was becoming widely accepted. I had hoped that scientists would be
allowed to work freely together across national boundaries, helping to
chart the course of future emissions reductions and assisting affected
communities in adapting to what lay ahead. But now I had seen that
the resistance to our findings was as strong as ever. The closer the
world came to acting on climate, the more likely it seemed that climate
scientists would come under attack for what we had found out.

There was nothing for it but to get back to work and search for
further evidence on the causes of climate change. And if many people,
fed a steady drip-drip-drip of misinformation by the *Daily Mail* and
others, thought that climate change was a distant and insignificant
issue, they might still be affected by the gathering tide of extreme
weather. Heatwaves, floods and droughts were on the rise.

Across the planet, weather is always changing. But in these early
years of the new millennium, weather records were being broken more
and more frequently. The human toll was mounting too, of people
seeing their homes destroyed and losing their lives in ever greater
numbers. Scientists needed to respond to this startling pace of change
and work quickly to better understand what was happening. I saw
an opportunity to show that climate change didn't just pose a deadly
threat for the future. It was a deadly threat that had already arrived.

5

In harm's way

My idea about the European heatwave of 2003 came to me while on holiday in Italy during that scorching record-breaking summer. It was early evening, just as the stultifying warmth of the day was starting to abate, when I had my moment of inspiration. My wife and I were sitting at a table on a restaurant terrace enjoying a glass of chilled wine as dusk settled over an alluring vista of the surrounding Tuscan countryside. All of a sudden, I saw a way to make a type of calculation that nobody had made before. Excitedly, I explained the idea to Pierrette. It was work intruding on holidays, but after ten years of marriage she had learnt to forbear when scientific inspiration struck.

We had come back to San Gimignano, the hilltop town where we had spent our honeymoon. Now, ten years on, we had found ourselves in the midst of Europe's most intense heatwave on record.[1] Luckily, we were healthy and able to adapt. We could make the most of early morning starts, take siestas during the middle of the day and enjoy the sultry evenings in outdoor restaurants.

Elsewhere, others were not so fortunate. Across Europe, over 70,000 people perished, many of them elderly living alone in sweltering attic flats in Paris and other urban centres like Zürich, Frankfurt and

London.[2] Hospital wards and corridors were packed with patients suffering the effects of heatstroke at a time when many doctors and nurses were away on their summer holidays. Many of the families of the vulnerable were also on vacation, often hundreds of miles away and unable to check on how their parents and grandparents were doing during night after night of sweltering temperatures. There had not been sufficient warnings about the dangers of the heat and the need to keep well hydrated nor had air-conditioned facilities been made available where the elderly most at risk could take refuge. Europeans were learning a hard lesson. Twenty-first-century societies in the rich Western world weren't prepared for weather like this.

After a week of creeping along baking Italian streets from one shady corner to another, Pierrette and I decided to decamp to the fresher mountain air of the Bernese Oberland. We stayed with my Swiss in-laws and went hiking. Our walks through alpine pastures would normally have been accompanied by the ringing sound of cow bells. But this summer, the hillsides were eerily quiet. Because the wells had dried up, the farmers had been forced to take their herds down into the valleys where they were now eating grass that should have been kept for winter feed.[3] It was yet another harmful impact of a disastrous summer whose effects were being felt far and wide.

When the weather finally broke and teeming rain gushed out of a black sky, I had more time to think about my idea. The key to it lay in the weather charts. It hadn't just been hot in Italy, France and Switzerland. Extraordinarily high temperatures stretched from Portugal in the west to the North German plain in the east, from the Mediterranean Sea in the south to the United Kingdom in the north. And until its final meteorological breakdown, this year's heatwave wasn't just a flash in the pan, here-today gone-tomorrow sort of weather event. It was extremely hot for months. The large-scale nature of the heatwave and its lengthy duration was what would allow me to do what nobody had ever done before – determine whether a single meteorological event could be linked to human-induced climate change.

For years, the scientific consensus had been that this wasn't possible, not because it was too difficult with current scientific tools, but that such a proposition simply didn't make sense. Up until then, all scientists were prepared to say was that a particular heatwave or flood might be consistent with the sort of weather you could expect more of in a warming world. We were not prepared to say greenhouse gas emissions were responsible. Weather was too complex for that, we thought. Random chaos made all sorts of extreme weather possible, with or without climate change.

Recently, however, the notion that it was impossible to attribute an extreme weather event to human influence had been overturned. In a commentary piece published in *Nature* in February 2003, Myles Allen proposed a new way of thinking that conceptually made attributing a weather event quite straightforward.[4] All you had to do, he argued, was think about such events from the point of view of risk.

To illustrate his idea, Myles told the story of a heavy rainstorm that had hit the neighbourhood of Oxford where he lived. Although his house narrowly escaped flooding, his neighbours were not so lucky and suffered the expense and mess of household inundation. The episode got Myles to thinking about whether anybody could be sued for some of the cost, in particular the oil companies who had profited handsomely from the greenhouse gas emissions that were warming the planet. If the risk of flooding in Oxford had been increased by those emissions, he wondered, shouldn't the polluter pay for the consequences of their actions? There were some complex legal issues to think about before such litigation could become a genuine possibility. But whether or not such a question ever finished up in court, the thought experiment got Myles to consider more deeply how blame could be attributed. This was a scientific question and one that was clearly tractable if you considered the example of weighted dice.

If you throw a normal die and you get a six, it's not that common but it isn't that unusual either. The probability is one in six, which means you can expect to get a six once in every six throws on average. If you do get a six, despite its comparative rarity, you can only attribute that throw to chance. But if the die has been weighted in such a way as to make a six twice as likely, that changes things. Now if you throw a six, you would have good reason to put some of the blame for the throw, not on chance, but on the nature of the die.

You can't blame the six you threw 100 per cent on the weighted die. But the weighting means that with repeated throws you would expect twice as many sixes as with an unweighted die. That means that half the blame of the six being thrown can be attributed to the weighting and not to chance. This is the analogy Myles drew for the attribution of extreme weather events.

If climate change had loaded the weather dice, Myles argued, you couldn't simply say that his flooded neighbours were unlucky. Were human activities to have changed the odds in favour of flooding then this was no natural disaster. Human-induced emissions *could* be held responsible, *if* climate change had changed the likelihood of flooding on that particular day in that particular place.

This is where it got tricky. In his article, Myles had not attempted to solve the knotty problem of how to actually calculate the odds of his neighbours being flooded. The scientific tools weren't currently there. British rainfall was too variable, the uncertainties associated with modelling too large, and the complexities of working out how a heavy downpour would affect a particular street of houses too great. Myles had raised a tantalizing possibility rather than a practical proposition. At the time his article appeared in early 2003, attributing an individual weather event to human activities still seemed little more than a distant pipe dream.

On holiday in Italy, it suddenly occurred to me that this pipe dream could become a reality, not for flooding in Oxford but for the current extreme heatwave that had spread its suffocating tentacles throughout much of Western Europe. The elevated temperatures were so large-scale and so long-lived that it was the sort of phenomena that current climate models *were* able to simulate well. Even better, Europe had the longest climate records in the world. There was a detailed temperature data set stretching back 500 years made up of thermometer readings, evidence from diaries and measurements from tree rings.[5] This would tell me not just about recent temperature trends but also, crucially, how temperatures could vary from one year to the next. All in all, I had all the jigsaw pieces I needed to put together the puzzle and work out how human-induced emissions had changed the odds of the 2003 European heatwave. That was the revelation that had come to me on that restaurant terrace in a baking hot Tuscany during the hottest summer the continent had seen in recorded meteorological history.

When I got back to my desk in England, I focused on one piece of work more intensely than I had ever done before. It had been like this towards the end of my doctoral research when I was working long hours to finish my thesis. But this time, the stakes felt even higher. If I didn't get a move on, I worried somebody else could beat me to it. Like many scientific breakthroughs, mine was based on putting two pre-existing ideas together, in this case the conceptual proposition of Myles's with my practical realization of how to use climate models and observations to make a calculation of changing weather odds. Like many scientific breakthroughs whose time has come, it seemed quite possible that somebody else would have the same idea too.

Day after day, I bent over my computer, writing and debugging the computer programs I needed to access the data and make my calculations. Desperate to get to a result, I would have worked without a break if I could but tiredness and fatigue is no friend of accuracy. To an outsider looking on perhaps, I was a normal scientist going about my normal day job, coming in to work each morning, taking breaks for drinks and lunch, leaving for home in the late afternoon. But inside, I was churning with impatience at my progress and excitement at my prospects. In a scientific lifetime, you tended to get very few shots at a career break like this. I didn't want to mess up my opportunity.

What I was attempting to do was work out the probability of temperatures peaking at the extraordinary values seen in 2003 averaged over the whole of Europe and taking summer as a whole. By pooling lots of data, I would be able to minimize the effects of random errors in the observations. I would also be able to compare what happened with a climate model and its simulations of European temperatures had we not changed climate by emitting greenhouse gases. Like comparing the rolling of two dice, one a normal unloaded die and another a tainted one that could potentially have been loaded to an unknown extent, I was aiming to calculate whether the probability of the heatwave was different, whether it had been changed due to human influence.

The model I used was the Hadley Centre climate model, HadCM3, the same model that had demonstrated so conclusively what had caused the twentieth-century rise in global temperatures. But while we had already shown that HadCM3 had an excellent simulation of temperature variations at the global scale, I also had to check it was equally good at simulating European temperatures. With grid boxes 250 kilometres across, I couldn't expect the model to be able to resolve the details of city centres like Paris or London. But I did need it to be able to accurately simulate variations in temperatures over Europe as a whole. Thankfully, my investigations showed that the HadCM3 model did reliably simulate temperatures on this continental scale. Considering the full summer season and such a large area would not only provide a more reliable result than zooming in on a few days at a particular place, it would also be more relevant to a heatwave that was so long-lasting and so widespread.

Completing research like this took months of work. During that time, thanks to my boss John Mitchell who recognized well when he needed to let his staff get on with concentrating on something important, I did little else during my working hours. Focused on one very specific investigation, I barely registered that elsewhere the world of climate science continued to move on. In particular, there was a new report from the Intergovernmental Panel on Climate Change (IPCC) to prepare.[6] One day, in March 2004, I was caught by surprise when my desk phone rang. It was a call that brought unexpected and significant news.

When I answered, it took me a moment to register that the speaker on the line was Susan Solomon, the eminent researcher who had made the breakthrough advance in explaining the causes of the ozone hole. Recently, she had been chosen to be one of the two co-chairs of the part of the next report from the IPCC that assessed the physical science basis of climate change, the 'Working Group I Contribution to the Fourth Assessment Report'. She wanted me to join her team of authors.

I listened to her explaining why, hardly believing that one of the most eminent scientists in my field had phoned to ask me personally. Her voice, with its weighty American seriousness, sounded impressive and what she was describing, the task of writing an IPCC report, onerous but important. She had selected me for my expertise in detecting and attributing climate change. As before, this part of the report was expected to come under the most intense scrutiny of all, providing as it did one of the most significant conclusions: the IPCC's latest statement on the causes of global warming. Susan had selected two leading experts to supervise the writing of the chapter in their role as coordinating lead authors. I knew them both from being members of the International Detection and Attribution Group together. One was Gabi Hegerl, who was known for her seminal work detecting the first signals of human-induced changes in surface temperatures, and the other was Canadian Francis Zwiers, known for his cutting-edge expertise in statistics. I would be one of a team of seven lead authors working under their overall direction. Together, we needed to provide the most comprehensive, robust and fully referenced assessment possible to support the crucial headline statements on the causes of climate change in the Summary for Policymakers document.

As I heard Susan say that it was going to be challenging but she had every confidence in the capability of the team she'd selected and the two scientists chosen to lead it, I realized that I needed to be ready with an answer when she asked if I would accept her invitation. Thinking rapidly about the implications, I knew this would be additional work on top of my own research and day-to-day responsibilities. But I would get to meet and collaborate with some of the world's leading climate scientists. And in doing so, I'd be helping to make our climate research matter for policymakers across the globe. When Susan's question came, my answer was immediate. I enthusiastically accepted her offer of a lead author position on the IPCC's Fourth Assessment Report.

For the rest of that day I didn't make much progress on my analysis of the 2003 heatwave. It was hard not to wonder about what being part of the IPCC might be like. I knew how difficult the second report had been for Ben Santer with the vicious accusations made by Patrick J. Michaels, Fred Singer and other climate deniers against him and how stressful the third report had been for my boss, John Mitchell. This fourth report was being prepared just as global efforts to reduce greenhouse gas emissions were starting to get serious. If Russia's game of diplomatic hardball resulted in them at long last signing up to the Kyoto Protocol, this landmark agreement would finally come into legal force. Now, more than ever, countries would need to be convinced that efforts to stop doing business as usual, to reverse decades of increasing emissions, were going to be worth it. The extent to which human activities could be blamed for rising temperatures, warming seas and melting ice was more relevant than ever before. Like Ben and John before me, I could expect to be involved in telling governments and citizens what some of them might not want to hear.

Having my research to focus on kept me from wasting too much time in idle speculation. What's more, if I wanted to publish my work on the European heatwave, I was going to have to make sense of some surprising results. I'd used data from the HadCM3 climate model and a mathematical technique called extreme value statistics to work out how often you'd expect to see the record-breaking heat of 2003. My results showed me that you could wait many hundreds of years before seeing such exceptionally high temperatures. This made the summer heat so unlikely it seemed hardly credible it had happened at all. Yet

I'd seen the measurements with my own eyes, felt the oven-like air on my own skin. What on earth was going on?

It was, I realized eventually, the galloping pace of climate change that I needed to take account of. In previous times, under a natural climate unaffected by human-induced emissions, the chances of such a summer were indeed extremely low. We didn't have weather stations thousands of years ago so we don't know for sure, but it looked like modern humans would rarely, if ever, have faced such an eventuality. We did know from the 500-year record of temperatures from tree rings and the 160-year record from thermometer measurements that nothing like the summer of 2003 had been seen before in those observations. And now, under a modest-sounding rise in global average temperature of just under 1 degree Celsius, the previous record had been smashed.

That gradual increase in average temperatures, I discovered, was having a striking effect on extreme temperatures. Their probability was increasing, not gradually like global temperatures, but at a rapid rate of knots. When I took account of human-induced climate change in my calculation of the probability of 2003-like temperatures, I found that they were now much more likely. And with more warming in future years, a repetition of such extraordinary heat would eventually become commonplace.

My headline result was that the probability of the record-breaking temperatures had increased about fourfold as a result of human-induced climate change. Nobody had ever published a figure like this before – of the changing odds of an extreme weather event due to climate change. Knowing this was a groundbreaking result, I eagerly set about writing up my methods and findings in an academic paper. It described how I had compared the odds of such extreme temperatures in a set of simulations of HadCM3 without human influence on climate with their odds in simulations that included human-induced emissions. Like a loaded die, today's world was much more likely to throw up the unusual towards which it had been weighted, in this case the terrible deadly heat of summer 2003.

Because of the difficulties inherent in estimating rare events from limited data and because I had to rely on a single climate model, albeit one that could accurately simulate past European temperatures, there was a large range of possible changes in likelihood consistent with the evidence I had available. But I was very confident that the probability of the 2003 heatwave had at least doubled. This was significant because a doubling of probability meant that most of the blame for the heatwave lay not with random bad luck but with the effects of human actions.

It was a striking conclusion and, on 21 May 2004, I submitted the results to *Nature* in a paper titled 'Human contribution to the European heatwave of 2003'. I did so with some trepidation because I knew I had only one shot at such a prestigious journal with this particular piece of work. I could probably get my results published elsewhere but that would delay their release and mean they would get much less attention. The risk of deadly heatwaves around the world was increasing rapidly. It felt vital that my findings were widely shared and soon.

In spring 2004, I felt blessed to have had the opportunity to work on the European heatwave. I'd had an idea in a flash of inspiration and I'd been able to produce results of sufficient importance to try sending to *Nature*. Although the event I was analysing had devastating and fatal consequences, I hoped my research would help improve the lot of vulnerable people like those who had been so hard hit in 2003. It felt well worthwhile to have demonstrated that the heatwave was no accident but instead was linked directly to human-induced greenhouse gas emissions. Now that governments knew that such temperatures were set to become a regular reality, it would encourage them to do more to help their citizens cope and work harder to cap emissions of greenhouse gases so the chances of events like this didn't keep increasing. Policymakers could only be grateful that the sort of research I was engaged in was coming to fruition.

Six weeks after submitting my paper, I was shaken out of my naïve complacency. In early July, I travelled to Moscow (see Chapter 4). My British colleagues and I thought it would be a relatively routine bilateral meeting between scientists from the two countries. Instead, I found myself a pawn in a geopolitical power game, a defendant in a show trial of climate science in which I and my ilk were accused of helping to wage war on Russia by Putin's right-hand man. I discovered at first hand what Ben Santer knew only too well from an American perspective: in the sphere of climate policy there were plenty of leading players keen to derail efforts to take our findings seriously. As far as I knew, nobody was trying to prevent me actually doing my research. But there were plenty of people intent on doing everything they could to diminish its apparent significance and relevance. As I saw in Moscow, they were prepared to go to great lengths – such as flying the world's leading climate deniers halfway round the world at a moment's notice – if they thought they could further the collective endeavour of climate change denial.

Returning home, I had a stronger sense than ever before that I needed to do what I could to help change the course of events on climate change for the better. Quietly beavering away helping to improve our scientific understanding no longer seemed sufficient. Thankfully, the chance had come along to work on the IPCC and help give the policymakers the best possible scientific advice. And if my paper was published, I wanted to talk about it to the media and tell the general public what I'd found out. That seemed the best way to make a difference.

*

In September 2004, I travelled to the first lead author meeting for the next IPCC report, which took place at the International Centre for Theoretical Physics in Trieste, Italy.[7] It wasn't the view of the sparkling Adriatic Sea from the grounds of the conference centre that excited me most when I arrived, but the heady buzz of anticipation in the foyer of the main meeting room. I was greeted by the two coordinating lead authors of my chapter, Gabi Hegerl and Francis Zwiers, and then it was time to take my place among my fellow climate scientists in the

Institute's old-style lecture theatre, where Susan Solomon delivered her opening address to the newly assembled team who had gathered from far and wide.

The international community was eagerly awaiting our report, our leader-in-chief told us. But ours was a job that could not be rushed. We were going to have to meet exacting standards. Every conclusion would need to be supported by evidence from peer-reviewed publications. Every finding would have to be carefully evaluated for its level of scientific confidence. Every summary statement would require a detailed argument to support it, grounded in the scientific literature. Fanciful speculation, although it might be the wellspring of much new scientific enquiry, was not what an IPCC assessment was based on. Our watchword was rigour. This is what was needed if we were going to meet the heavy weight of expectation placed upon us by the world's governments and citizens.

I felt inspired by Susan's words. Her prescription for how we should write our report was based on a passionately held conviction in the need for getting things right. As a mathematician by training I found this adherence to rigour deeply attractive. Rigour would ensure we provided the best scientific advice to governments and rigour would defend us from climate deniers' attacks. I felt privileged to be sitting there as part of this international team of experts tasked with advising the world on climate change.

After the coffee break, each of the eleven teams for the different chapters were ushered across to their allocated rooms throughout the complex so that they could begin their deliberations. Between us, we covered the full range of climate science, from observations to models, from past changes to future projections. The task we faced in our chapter was to make an assessment of the causes of climate change. This was the topic that would form the crucial centrepiece of the whole assessment, the basic piece of information motivating action on reducing greenhouse gas emissions, the answer to why climate was changing. To get there, we had to start marshalling the scientific evidence from peer-reviewed published papers and plan out over the next few months the writing of the first iteration of our chapter, what we called the zero-order draft.

We took our seats in a seminar room in one of the Institute's upper floors. In addition to Gabi, Francis and me, the group was made up of Nathan Gillett, who had recently completed a PhD under Myles Allen in Oxford, Neville Nicholls, a distinguished researcher from the Australian Bureau of Meteorology, French climatologist Pascale Braconnot, Yong Luo from the China Meteorological Administration, American climate scientist Joyce Penner, and Jose Marengo from the Brazilian Centre for Weather Forecasting and Climate Studies.[8] To kick things off, we were all asked to make a short presentation outlining what we considered to be the main emerging pieces of research. At long last, as we settled down to listen to what each of us had to say, the work on writing the next international climate assessment had begun.

My turn to present came in the first session of the afternoon. I had thought that I was going to spring a surprise on everybody by describing my work on the European heatwave. Given it was still under consideration by a journal and not yet common knowledge, my findings would be new to most people here. But, as it turned out, my two coordinating lead authors were ready to spring a surprise on me. My work was already undergoing a rigorous process of anonymous in-depth review by experts appointed by *Nature*. But Gabi and Francis wanted it to undergo another rigorous round of scrutiny by the rest of my chapter team.

How had I calculated the degree of warming seen in Europe that could be blamed on human rather than natural activities, they wanted to know? What form of extreme value statistics had I used to calculate the odds of temperatures reaching the extreme values seen in 2003? How had I checked the reliability of HadCM3 in simulating the variability of European temperatures?

During a lengthy grilling I did my best to explain all the ins and outs of exactly what I had done. It was an uncomfortable experience.

If anybody could find a weakness in what I'd done this crowd would, and this was at a time when I was on tenterhooks to hear whether my research was sufficiently strong to justify a billing in science's most prestigious academic journal. It reminded me that, with friends and colleagues like these, there was nothing I could hide scientifically. This was exactly as it should be if collectively we were going to judge the merits and limitations of the scientific literature we had to assess. In any event, as I reflected later, if my research had survived this interrogation unscathed, there was a good chance it would stand up later too, not just with the reviewers who were examining my submission to *Nature*, but also with the wider scientific community when they got to read the paper and consider what I'd done.

The encounter was a perfect illustration of what Susan had talked about earlier when she impressed on us the need for rigour in our assessment. In practice this meant understanding every last detail of key scientific findings, their strengths and weaknesses. This was going to take a significant amount of time and energy. Collectively, we would have to weigh up the evidence in front of us and the associated scientific uncertainties and, for the final report, agree on the level of confidence that should be placed on our summary statements. At this first lead author meeting in Trieste we had only just begun this laborious process of assessment. And in her final summing up before we left the Adriatic coast, Susan reminded us that the process of critical review of our IPCC report had only just begun.[9] Whatever judgement we reached on the work of scientists inside or outside our chapter team, it would have to survive the critical scrutiny of the scientific community at large, not just once but over the course of three rounds of review of successive drafts of the report.

Returning home, I set to work developing the section of our chapter on temperature changes. I read the papers that had recently been published and the preprints of papers currently under review that we were aware of and started to summarize the current understanding. A key question was exactly how much of the global average warming over the past fifty years could be attributed to greenhouse gas increases. With plenty of time left for further research before our report was finalized, I wasn't yet aware of all the literature that we would eventually be able to cite. But the papers I did have access to pointed to greenhouse gases being the dominant cause of warming over the last half-century.

While I worked on the report, I waited anxiously for news from *Nature*. At last, on 5 October, word came through that my paper had been accepted after two rounds of review by independent experts. It was a relief. A year's hard work had been rewarded with publication in a high-profile journal. But success brought with it a nerve-wracking prospect. *Nature* worked hard to ensure that its papers were widely publicized. I had the opportunity I'd hoped for, to get this new science talked about and help motivate action for the good of everyone threatened by climate change.

Such a chance didn't come along very often. I had to try to make the most of it. My paper would be included in the 2 December issue. I had a few short weeks to gather my thoughts as to what I wanted to say when the media came calling.

During the week leading up to publication I found myself much in demand for comment. The journal had already sent out a press release and reporters had an opportunity to talk to me under embargo in advance of publication. The release outlined my main conclusion that 'Human influence has at least doubled the risk of a regional heatwave like the European summer of 2003'.[10] Most journalists understood the concept of changing odds and quickly appreciated the main aspects of my analysis. Explaining how I'd made my calculations seemed to be relatively straightforward. But explaining the relevance of my findings turned out to be trickier than I had expected.

The press release had included a quote from Swiss and German experts that my study 'might profoundly affect the course of international negotiations on climate change mitigation'.[11] Although it seemed a strong-sounding statement, many journalists didn't see how my results would have such a dramatic effect on climate policy.

To them, the urgency of the climate issue did not seen so apparent. I tried suggesting the potential for legal actions of the sort Myles had written about when his kitchen was nearly flooded. But to many interviewers, this idea seemed rather far-fetched. Barristers arguing over the ins and outs of scientific research didn't sound like the most obvious way of solving the climate problem.

And then I mentioned another finding from my paper, even though at the time I thought it secondary to my main result. It was one that didn't even appear in *Nature*'s press release. As well as the increased risk of the 2003 heatwave, I had calculated the risk of European heatwaves in future. I had found that by the 2040s, one summer in two could be warmer than 2003 if atmospheric greenhouse gas concentrations continued to rise. To me, this wasn't that surprising. It was just a direct consequence of the fact that global warming dramatically increases the frequency of heatwaves. But to the journalists I spoke to, this was extraordinary. Without action to reduce greenhouse gas emissions, the heatwave summer of 2003 looked set to become a regular occurrence in just a few decades. That, they found, was very interesting indeed.

My result added to the growing evidence that climate change was bad news. Scientists working on an early draft of the IPCC's next report on the impacts of climate change – the Working Group II Contribution to the Fourth Assessment Report – could point to a wealth of findings outlining the likely future effects of more extreme weather. An increasing frequency of damaging floods would bring misery to millions of people in the mega-deltas of Asia, the low-lying islands in the Pacific and other coastal regions of the world. More intense droughts would shatter cereal productivity across the lower latitudes, dramatically increasing the risk of starvation. And increasingly violent storms would destroy homes, lives and livelihoods, and raise ill health from malnutrition, diarrhoea and infectious diseases. The earth's ecosystems too were at peril. Wildfires would become more common. Many plant and animal species would be at increasing risk of extinction. Coral reefs in the ocean would be lucky to survive.[12]

All in all, it was a grim prognosis, one that should spur collective actions to reduce emissions. But these predictions were based on approximate estimates of how future extreme weather would affect people and wildlife. In the same way that climate denier Fred Singer had lobbied at the 1997 Kyoto climate negotiations, such prognoses could be attacked as being hypothetical. What journalists found most noteworthy about *my* findings, it seemed, was that there was nothing hypothetical about a summer that had killed 70,000 people. By demonstrating how that terrible toll was linked with human-induced greenhouse gas emissions, I had brought a hazy future under climate change into much sharper focus. No one could dispute the disaster that was the European heatwave of 2003. Clearly, yet more warming in future held stark implications for human safety.

The next day I eagerly scanned the news-stands to see what the papers had printed about my work. 'Summer of scorching heat will be back' was the headline in *The Times*.[13] 'Deadly hot summers to become the norm', was the *Independent*'s take.[14] 'Extra scorchio', said the *Daily Mirror*'s headline, above a cartoon of a sizzling sun and a sub-heading which read, 'climate change to put last summer in shade'.[15] And it wasn't just the British press taking an interest in my findings. *Le Monde*'s headline that summers were becoming hotter and hotter sat above an article describing how the sort of heatwave that had killed 15,000 people in France would occur one summer in two by 2040.[16] The German paper *Der Spiegel* described a man-made heatwave that had killed around 7,000 in Germany. By showing that a recent weather catastrophe was a foretaste of the future, an expert from the German Weather Service called my work 'a breakthrough'.[17]

If the European press were on the whole supportive of my work, and keen to hail it as a major advance, the papers in the United States carried a different perspective. The *Washington Post* quoted Myron Ebell, director of global warming and international environmental policy at the climate change denial lobby group, the Competitive Enterprise Institute. 'This is a very small-potatoes paper,' he said. My results were 'based on modelling that can't be proved or disproved for the next 50 years. Modelling is not science.'[18]

There was also more interest in the question of litigation on the American side of the Atlantic. In the same *Washington Post* report, Annie Petsonk, international counsel for an organization called Environmental Defense Fund, thought that results such as mine could cause investors to re-evaluate where they put their money. Just as they had pulled out of the tobacco industry when the link was established between cigarettes and cancer, Petsonk wanted industry to face up to the 'potential global-warming liability that companies that refuse to limit their greenhouse gas emissions will face'. To her, my paper marked 'the first time scientists have demonstrated the human fingerprint on a particular weather event with the kind of certainty that will stand up in court'.

Thankfully, whichever side of the Atlantic I looked, it seemed like my paper had hit a nerve. On one side, it sounded a warning call to politicians by demonstrating the perils of delay if international negotiations failed to reduce emissions. On the other, it provided future ammunition for lawyers in their efforts to steer us towards a greener future.

I found it easier to see how my research fitted into the European perspective. Governments now had an additional piece of ammunition in their fight to build agreements to reduce emissions, one that would resonate particularly with those citizens affected by the 2003 heatwave. The American perspective felt more problematic. Mounting successful litigation on the basis of emergent science against fossil fuel interests supported by adept and well-funded legal teams seemed a perilous undertaking. I didn't fancy being an expert witness in a court of law. I imagined having to defend myself from attack by cunning lawyers, the sort of twisted assault with backing from climate deniers I'd experienced in Moscow. Thankfully, my appearing in court seemed a rather distant prospect, one for American colleagues to worry about rather than me.

Instead, I had the next IPCC report to concentrate on. In a few months, in May 2005, I would be attending the second lead author meeting in Beijing, where we would start to develop the first order draft, the first version of our report to be sent out to widespread expert review. We would reconvene for a third lead author meeting, in New Zealand in December 2005, when we would work on an improved second order draft in response to thousands of critical comments. That draft would be sent out for review again and in our final meeting in Norway in June 2006, we would prepare a final version of the report and the Summary for Policymakers to put before governments at a final meeting. That crucial gathering would take place in early 2007 in Paris, the very city that had suffered the worst effects of the 2003 heatwave.

6

Very likely due

The meeting that took place in Paris from 29 January to 1 February 2007 was the most significant meeting about climate science yet held. It was there that scientists and government representatives would approve the Working Group I Contribution to the Fourth Assessment Report of the Intergovernmental Panel on Climate Change (IPCC). This would provide the most definitive scientific view on the nature and causes of global warming. It followed previous IPCC reports, the most recent of which had been published in 2001, that had helped drive international climate negotiations to reduce greenhouse gas emissions. But this new report was of particular significance.

The Kyoto Protocol, signed in 1997, had finally come into legal force with its ratification by Russia in November 2004 after the EU dropped objections to them joining the World Trade Organization.[1] But Kyoto didn't do nearly enough to address the global threat of climate change. It was only a first tentative step by some industrialized countries. Further international negotiations were needed if the world was going to take the next much more significant step of agreeing the substantial global cuts in emissions that were needed. These negotiations were due to culminate in a crunch summit in Copenhagen in November 2009.

For the negotiators to make progress they needed an appropriately robust set of statements by the IPCC summarizing the latest science.

It all hinged on four long days and nights in the Headquarters building of UNESCO not far from the Eiffel Tower. It was in the building's bunker-like congress centre that hundreds of government representatives and scientists would debate every single word of the report's Summary for Policymakers. This would then be released to the world's media at a press conference scheduled for 10 a.m. on the fifth day of that fateful week, Friday 2 February 2007.

Not all of the authors of the report were invited to Paris. All of the coordinating lead authors had come, including Gabi Hegerl and Francis Zwiers, as had two other authors from each of the chapters. I had been invited because of my expertise in the statistical basis of our conclusions on the causes of climate change, and Australian climatologist Neville Nicholls was also invited because of his expertise in the observational basis. Collectively, the task of the scientists representing all the different chapters was to convince governments of the correctness of our overall assessment. If we succeeded, the negotiators in Copenhagen would have a basis on which to act. It was an exciting prospect helping shepherd the report on which we had worked for so long to a successful conclusion. But it was also a heavy responsibility, given what was at stake.

Over the four days set aside, there were hundreds of sentences to consider, each of which had to be agreed by every single government in consensus. If a government wanted to suggest a change of wording, it would only be implemented if the scientists agreed to it. If not, we could propose a better alternative or else reject the change entirely. But governments too could object to what we proposed. If they did, we would have to try to convince them of the strength of scientific evidence that supported our proposal.

Two statements were likely to attract the most attention. One summarized the state of knowledge on the observations of global warming. The other summarized the state of knowledge on the causes of global warming. Together, they provided the motivation for why countries should step up efforts to reduce emissions. I would have to help defend the statement on causes of warming. Knowing the history of the IPCC, including the attempts by Saudi Arabia and Kuwait to undermine the report when Ben Santer was involved a decade before, what awaited me felt like a daunting prospect.

The hall where history would be made could accommodate the hundreds of people involved, but only by packing them together at long rows of tables in an environment that would become more and more claustrophobic as the debate intensified. Government delegates were assigned their places in alphabetical order from Australia and Austria at the front to the UK and USA at the back. In front of them was a raised platform from which the report's author-in-chief, Susan Solomon, chaired the meeting alongside her Chinese co-chair Qin Dahe. As each section of the summary document came up for discussion, they would be joined on the platform by the relevant authors to defend their part of the assessment.

At the start, progress through the document was painfully slow.[2] During the first three-hour half-day session – and with only four days to approve the whole seventeen-page document – the sum total of three sentences was approved. Countries proposed minor changes to wording or major changes to organization which were either unnecessary or undesirable. Susan was stubborn and determined not to be waylaid. At this early stage it was crucial she took time to establish her authority. It would be needed later when we got to the more important and contentious sentences in the report.

By the first afternoon, the pace had picked up a little. Even so, thanks to a continuing desire from delegates to request minor edits and ask for further clarifications from the authors, by close of play only half a page had been approved. Tuesday continued in the same slow vein as we ploughed laboriously through the section on observations of climate change while Susan resisted attempts by delegates to make unwarranted changes to the text. It meant long hours on the podium for the authors who had joined her there, including University of East

Anglia scientist, Phil Jones, the foremost international authority on global surface temperature records. For the rest of us, it meant long hours watching on, being prepared to intervene on points of detail if called upon from the chair. Otherwise all we could do was wait for our big moment on the podium when our own section would be under debate.

By late on Wednesday evening all the sentences in the section on observations had at last been gavelled down. Except for one – the first of the two crucial statements that were likely to attract the most public attention later. If, that is, they were agreed to by every single government in the room.

'Warming of the climate system is unequivocal', it stated, 'as is now evident from observations of increases in global average air and ocean temperatures, widespread melting of snow and ice, and rising global average sea level.' As the words appeared on the giant screens at the front of the hall, the distracted busy-ness of the preceding hours of technical debate – the rustling of papers, the whispered conversations, the footsteps of participants going to and from the foyer – was suddenly damped down to be replaced by a hushed air of concentration. Sitting patiently in the row of seats reserved for authors not directly involved in the discussion from the podium, I had wondered for months how this choice of words would fare, ever since the phrase had been agreed at our final lead author meeting six months before. It was finally time to find out.

For us scientists, the choice of the word 'unequivocal' was one on which we weren't prepared to compromise. The evidence for global warming from observations of the atmosphere, the ocean and the cryosphere was utterly undeniable. There were inevitably small errors in temperature measurements made over the years, but these errors were completely swamped by the overwhelming consistency of different types of data. Whether they be from weather stations, satellites or ocean buoys, they all pointed to a warming climate. For years, the climate deniers had claimed there was still some debate about the reality of global warming. With the latest measurements all pointing one way, towards a warming world, that debate was now over. As authors of the report, we were adamant that the IPCC needed to say so. It was vital that governments acted on the science as it was, which proved that global warming was a reality.

The interventions from the floor made it clear that not all delegates were going to readily agree. One of the governments proposed that we should use the word 'evident' instead. Phil Jones and his colleagues on the podium firmly resisted, carefully explaining why this alternative didn't express the unequivocal reality of global warming. Several other nations spoke out in favour of the authors' choice of 'unequivocal'. Judging by their support, it looked like this might be the most popular choice. But without universal agreement, Susan couldn't record the sentence as having been approved. We needed consensus.

Even as the clock ticked on towards midnight on the third of the four days with two thirds of the document still to be considered, Susan knew she had to give this crucial sentence time. If consensus proved impossible, there was an option to insert a footnote stating that one or more governments didn't agree. But such a device was hardly ever used by the IPCC. Governments rarely want to go out on a limb by distancing themselves from a scientific statement that has been accepted by everybody else as correct, and scientists are very reluctant to accept a weakening of their document through it not having been agreed unanimously. As I waited to see what would happen next, I fervently shared in my colleagues' strong attachment to the 'unequivocal' word. All of us wanted to avoid such a footnote if at all possible.

At this moment of crisis, Susan managed the room with aplomb. One after another, she asked governments who had raised their flags and she knew to be on our side to make an intervention. Sensing their chance to be supportive, country after country spoke strongly in our favour. Together, they built up a case that became impossible to resist, diplomatically as well as scientifically. Eventually, seeing no further objections, Susan gavelled down 'unequivocal'. Phil Jones and his colleagues stepped down from the podium, looking relieved and shattered after almost three days defending their proposals. Everyone on the author team relaxed. The meeting's first big moment of high stakes had passed off successfully.

The next big moment of high stakes came sooner than I'd expected, on the morning of the following day, the meeting's last. With a large chunk of the document still to be approved, Susan decided that our section on the cause of climate change would be the first item on the main agenda of the day. There were still parts of sections that came before ours and which hadn't yet been discussed. But agree our crucial statements in the morning, she was calculating, and the mood in the hall in the afternoon could change from anxious about the snail's pace of progress to much more upbeat about our chances of finishing by the end of the day. The rhythm would pick up. There would be a fighting chance that the rest of the document, including its long section on projections of future changes in climate, could be approved in time for the press conference the following morning.

I felt a sudden surge of adrenaline as I took my place on the stage at the front of the room alongside Gabi Hegerl, Francis Zwiers and Neville Nicholls. For the first time this week, I was putting myself firmly in the spotlight. Many of the delegates in the hall in front of me were looking up from their laptops and scrutinizing the four of us, this new set of arrivals to join Susan on the podium. What did they make of us, I wondered? And what did they make of our summary of the causes of climate change that we had spent the last three years preparing?

It had taken us four lead author meetings and countless hours dealing with thousands of review comments to finally thrash out our definitive conclusions. We thought we had provided the best possible scientific advice for politicians to take into future climate negotiations because we had firmly based it on a thorough assessment of current scientific understanding. But the tenor of that crucial, impartial advice depended entirely on what happened over the next couple of hours. I took a deep breath and tried to remain focused on what I knew best: the detailed chain of arguments that linked the latest body of peer-reviewed scientific literature to our summary statements about to be put to delegates.

Susan introduced the first sentence up for debate, which had also appeared on the large screen behind us.[3] There were eleven short paragraphs of sentences like this to present. They provided a concise summation of the many different lines of evidence we had to draw on. It was necessary to provide the governments with the strength of our findings. But it was also a lot to get through. Those of us at the front could only hope that on this final day of the meeting, delegates were prepared to up the pace of approval. If not, and we hadn't finished our section by lunchtime, our chances of finishing the whole document on time would lie in tatters. The most important of the sentences – the one that summed up our findings in a stand-alone headline statement – would be dealt with last. This would be the vital crux move in this morning's make or break approval session.

As we waited to see if there would be an intervention on the first sentence Susan had read out, I hardly dared breathe. Imagining the people in front of me wondering if this new face could be trusted as a reliable witness, I didn't want to shift uncomfortably in my seat or exude anything other than a Zen-like calm. Churning inwardly, the next hour or two felt momentous. A successful outcome would depend on the scientific groundwork that we had collectively undertaken as authors, homework we would need to call on if any of our statements met with stiff resistance from the floor.

Over the last three years, we had spent many hours discussing the findings of scientific papers and choosing appropriate language that expressed the confidence we could have in our conclusions. We had stronger evidence than ever before that the warming of the atmosphere and oceans, the rising sea levels and the diminishing snow and ice could only be explained by human activities. Our assessment reflected this. But we had also taken account of some remaining difficulties in monitoring changes in climate and in pinning down exactly how much global warming could be attributed to greenhouse gas emissions.

Difficulties in calculating this attributable warming had caused a stumbling block in the approval of the previous IPCC report in Shanghai six years before. When the Saudi delegation objected to the term 'substantial' to describe the amount of global warming due to greenhouse gases, John Mitchell and his author team found it hard to find suitable alternative wording. There was an absurdity about global efforts to halt global warming depending on governments and scientists agonizing over the meaning of words like substantial. But it was also the case that we couldn't afford to stray from the scientific evidence in our choice of words, else we would risk undermining the whole motivation for climate action. That had to be based on a careful account of the facts, not on the sort of fake news that climate deniers would have governments follow.

Thankfully, we now had much more information based on a greater wealth of scientific studies. And it was with this crucial issue that the serious business of building the foundations of our assessment had begun, at our first lead author meeting to prepare the fourth IPCC report, two and a half years before by the side of the Adriatic Sea in Trieste.

*

Back in September 2004, after presenting my work on the European heatwave to my chapter team and facing a grilling from my fellow authors, I went to meet another of the chapter teams, the group who were working on the section on future changes in climate. They wanted to know more about identifying the characteristic fingerprints of natural and human influences on climate. They were keen to hear from me because of the work I had done with my colleague from Oxford University, Myles Allen, which showed that quantifying the degree of past global warming to greenhouse gas emissions also helped with predicting the rate of future warming.[4]

The expected way in which surface temperatures responded to increasing greenhouse gases, I explained, differed markedly from how it varied naturally. This meant it was possible to identify the distinctive fingerprint of human-induced climate change, in which the land warms up faster than the ocean and the Arctic warms up fastest of all. This helped us to work out how much of the observed warming could be explained by human factors and how much could be explained naturally. But there was a catch.

As well as adding to the greenhouse effect, burning of fossil fuels leads to air pollution and small particles in the atmosphere – so-called aerosols – that reflect sunlight and cool climate. This cooling effect complicated our task in attributing past warming to human factors. The observed global warming could have been caused by a relatively large amount of greenhouse-induced warming counteracted by a relatively large cooling from aerosols. Or the same warming could have resulted from rather less greenhouse warming and rather less greenhouse cooling. To try to work out which it was, we had to use our fingerprinting techniques.

Our fingerprinting techniques were capable – just about – of distinguishing the cooling effects of aerosols from the warming effects of greenhouse gases.[5] But, like a dolphin spotter looking through not-quite-powerful-enough binoculars while trying to establish individuals in a pod by their dorsal fins, distinguishing the atmospheric fingerprint of greenhouse warming from the fingerprint of aerosol cooling was proving a struggle. We had limited information to go on.

The best feature to distinguish between the effects of greenhouse gases and aerosols was the one that Ben Santer had raised in that explosive session at the American Geophysical Union with Patrick J. Michaels during my 1996 initiation into the way deniers funded by lobby groups attack climate science.[6] As Ben explained then, aerosols cooled the northern hemisphere more than the southern hemisphere, particularly during the 1950s and '60s when air pollution was increasing rapidly. This difference was sufficient to give a rough idea of how much of the world's temperature change was due to greenhouse gases and how much to aerosol cooling. But with the fingerprinting technology of the time, difficulties still remained.

Our calculations showed that greenhouse warming over the twentieth century was somewhere between 0.6 degrees Celsius and 1.3 degrees Celsius and the cooling due to aerosols between 0.1 and 0.7 degrees Celsius.[7] The difference between the warming and cooling explained the net temperature increase of about 0.7 degrees Celsius. The net result was that greenhouse gases most likely caused more global warming than had actually been observed. This was significant because it meant that human activities had caused even more damage than might have been supposed by looking solely at the climate records. It was a finding with important but depressing implications.

With our not-quite-powerful-enough fingerprinting binoculars, it was hard to be very precise about the size of this aerosol effect. What we did know was that aerosols, by cooling the planet, had reduced the actual warming somewhat and this should be factored into projections of future warming. If those aerosols were removed in future by cleaning up air pollution, there would be an unfortunate side effect for the planet. Future warming would be even greater than that expected from greenhouse gas increases because of the elimination of this additional aerosol cooling.

This was heavy stuff to describe and on this particular day I was pleased when we could break for lunch and sit outside in the sunshine while watching the odd yacht trundling in and out of port. Explaining our fingerprinting approach to scientists not familiar with this work was challenging. There were sound physical principles involved although it could still seem mysterious to some people. But we had shown it worked. The well-tested method, invented by Myles Allen and unveiled at that 1996 San Francisco meeting with Ben and Michaels, had subsequently been published and applied to a range of different climate models.[8] The results from several peer-reviewed studies were all consistent in finding that greenhouse gases played the dominant role in past warming.

Even so, at the time of the first lead author meeting in Trieste in September 2004, I was thinking about how best to improve those fingerprinting binoculars. Several of us, including Myles and a colleague from the Centre for Ecology and Hydrology called Chris Huntingford, had started to develop a more sophisticated way of distinguishing fingerprints by using the results of a range of climate models, all of which were providing important information but all of which, like any scientific model, could have errors in its simulation of the variables we were trying to pin down.

This new method, which we called 'Error in Variables' or EIV for short, was ready to submit for publication about a year later, on 13 October 2005.[9] At the time, I thought it a useful but relatively minor step forward in the hunt for more and more accurate attribution of climate change to natural and human factors.

I didn't foresee its crucial significance for the IPCC report when I revised our paper on EIV as it went through two rounds of review before being published on 14 March 2006, nor did I imagine its later importance when we included it in our chapter's assessment. Not even when I was on the podium at UNESCO HQ with Susan, Gabi, Francis and Neville, waiting anxiously for the first intervention on the first sentence of the section on understanding and attributing climate change of the Summary for Policymakers, did I predict the pivotal role it was going to play. But that was one of the delights of working on the IPCC. There were always surprises.

After the previous day's approval of the 'unequivocal warming' statement, a refreshed and purposeful body of delegates caused little trouble as our sentences were placed one by one on the big screen. These sentences were the work of many hours' deliberations in our author team and covered a myriad of different aspects of the climate system and how their changes could be ascribed to natural and human-induced factors. They described how human actions had contributed to increasing ocean temperatures, rising sea level and reducing sea ice. They found anthropogenic not natural explanations for a greater risk of heatwaves and changes in the track of storms. They explained how the observed patterns of warming, including greater warming over land than ocean and their changes over time, could only be accounted for by human-induced climate change.

For some sentences, a delegate proposed a minor rewording for clarity, most of which we accepted if we thought it improved the readability or made our intentions clearer. For others, if a country representative asked a question about the basis of a statement, Gabi or Francis were able to provide a brief clarification that was sufficient for the sentence to pass. While the document couldn't pass without government approval, no changes, we knew, could be made without our agreement. We scientists, not the governments, held the pen.

One after another, our sentences were quickly accepted. It looked like approving our conclusions on attribution was working out to be rather straightforward. I began to feel a bit less tense. I dared to imagine relaxing later, looking forward to sometime tonight when all the other sections had been completed and we could venture outside for a celebratory drink. But we were not done yet, not until we had considered our headline statement. And approving that, as it turned out, proved to be far from straightforward.

The sentence in question stated: 'It is very likely that anthropogenic greenhouse gas increases caused most of the observed increase in globally averaged temperatures since the mid-20th century.'[10] The term 'very likely' was an example of the IPCC's use of special terms which had already been defined earlier in the document. It meant that we assessed that there was at least a 90 per cent chance of the statement being correct. The previous IPCC report that John Mitchell had seen approved in Shanghai had concluded that it was 'likely' that increased greenhouse gas concentrations caused most of the warming. 'Likely' in IPCC terminology meant a greater than 66 per cent chance. In going from 'likely' to 'very likely' in our headline attribution statement, we were making a clear statement that the strength of evidence for human influence on climate had increased significantly over the previous six years.

As soon as it was placed on the big screen, the Chinese delegation turned the name board of their country to the vertical indicating that they wanted to make an intervention.

Their country was a rapidly expanding economy whose emissions of greenhouse gas were also rising rapidly. They stood to lose substantially if reducing those emissions constrained their economic growth. At the same time, the many threats from climate change, including the risks of flood and famine – in China as elsewhere in the world – would grow alarmingly if China's emissions continued to grow unchecked. In tackling the climate crisis, the world's most populous country was also one of its most important. At this meeting, as in many other diplomatic occasions worldwide, when China spoke, everybody else listened with great attention.

Called to speak by Susan, their representative set out in a short speech the nature of his country's objection. It was a basic one. The science had not advanced sufficiently, he maintained, to justify a stronger statement this time than had been agreed last time in Shanghai.

I could feel my pulse racing as I contemplated what had just happened. I had half expected problems from Saudi Arabia, the nation with a known track record in trying to block or water down IPCC statements. But this was China, one of the most powerful nations on earth, announcing that they could not accept the statement that was potentially the most vital in the whole report.

I had no idea what would happen next or what if anything I could do to help. All I knew was that this was the one sentence towards which all our collective effort in our chapter had been building these last three years. This is what I imagined being headline news tomorrow, a summary of the scientific consensus that would be quoted for years to come. And now its passage dangled by a thread. China had very publicly and very demonstrably announced they couldn't accept it. How on earth, I wondered, could they back down from that?

Susan invited the two coordinating lead authors of our chapter, Gabi and Francis, to explain our reasoning. The basis for our increased confidence was set out in detail in our chapter of the report, Francis explained, as calmly as he could. As, between them, they set out to boil down the thousands of words that explained the reasoning behind this summary statement into a concise and persuasive case, I could hear the barely disguised anxiety in each of their voices. At the time of the previous IPCC report, Gabi went on, the assessment was based largely on global temperature changes at the surface. Now we had evidence of the effects of human activities on other aspects of the climate system, including warming below the surface of the ocean, warming on different continents, reductions in Arctic ice and changes in global wind patterns. These findings provided additional confirmation, not available six years before, that climate was changing due to human actions.

Following these remarks, Susan took a host of interventions from many of the other governments present. Thankfully they were all supportive of the statement that we scientists had proposed. Many of them reflected approvingly the argument we had put to them, that the growing evidence of consistency across a changing climate increased the confidence in attribution of climate change to human influence. I hoped that would change China's tune. But when Susan came back to them, their delegate was sticking to his guns.

Regarding changes across the climate system, he insisted, there remained many scientific uncertainties which meant that such evidence did not strengthen the attribution statement. Listening to him talking, I realized he was trying to make negotiating capital out of the fact that it was easier to measure changes in climate at the surface than elsewhere. Over land, we had accurate thermometer readings from weather stations. Over sea, we had reliable measurements of the temperature of seawater at the surface taken from ships. But I was painfully aware that away from the surface, taking measurements became more difficult. This included taking the temperature of the deep ocean.

*

The evidence of ocean warming provided one of the most important confirmations of the hypothesis that global warming was mainly due to greenhouse gas emissions. Climate models predicted that over 90 per cent of the additional heat trapped in the climate system by the enhanced greenhouse effect should go into warming up seawater. This was exactly what the data showed then and what they still show today. But at that time, in January 2007, those same data also threw us a curveball, one that did have some effect on the confidence we could have in our attribution statement.

During the 1980s, the measurements appeared to show the ocean *cooling* after a period of accelerated warming in the 1970s.[11] Warming of the deep seas resumed in the 1990s and had continued ever since. Robust attribution of global warming to human factors depended on ruling out natural variations as an explanation. Based on surface data, we could do that. But if the sub-surface ocean data were right, the deep seas could cool naturally to a much greater extent than predicted by climate models. This could mean the oceans were subject to large natural upheavals. If so, that raised the possibility that such upheavals could have caused at least some of the warming observed at the surface.

In our IPCC author team, we had spent many hours over the last two and a half years discussing this issue. To help solve the puzzle, we had turned to colleagues from the chapter on observations of ocean climate change who were experts in oceanography and well versed in the difficulties of measuring temperatures under water. Over the course of four lead author meetings and countless email exchanges, our conversations with the oceanographers helped us figure out what such difficulties meant. It all hinged on the rate at which an instrument probe falls through water when thrown over the side of a ship.

It has often been said that we know more about the surface of Mars than the interior of our planet's oceans. But since the start of the twenty-first century we have had a much clearer picture, thanks to thousands of automated buoys, called Argo floats. These yellow, six-feet-tall, torpedo-shaped automata drift around in the currents before rising to the surface every ten days to beam their readings of temperature and salinity back to watching satellites. Their transmissions tell a picture entirely consistent with the predictions of climate models, of a steadily warming ocean.

Before Argo, oceanographers had to rely on measurements taken by launching buoys into the sea and by taking readings from research ships. During the 1970s many of these readings were made using an instrument called an expendable bathythermograph – or XBT for short – a small probe which was jettisoned overboard and whose measurements of undersea temperatures were relayed up to the surface along a copper wire. The probes had no depth recorders so the locations of their measurements had to be figured out from the rate at which they fell through the water. An accurate record of ocean temperatures depended on an accurate knowledge of the fall speed of the XBTs.

Unfortunately, it turned out, scientists didn't know the fall speed very accurately. To make matters worse, the oceanographers told us, changes in manufacturing of these XBTs at the factory meant that their fall speed was changing over time. In particular, during the late 1970s and early 1980s, their fall speed probably slowed down substantially. As a result, measurements thought to have been taken at particular depths were actually being taken at much shallower depths because the XBTs were not falling as quickly through the water as had been assumed. As temperatures reduce with depth, this meant ocean temperature data sets were probably showing readings that were too warm during the late 1970s and early 1980s. The apparent warming spike in ocean temperatures during this period could be a result of mistaken assumptions about the fall speed of XBTs. But we couldn't be certain, not until there had been a complete re-evaluation of those XBT measurements in a brand-new ocean temperature data set.

In the meantime, we had to come to a view of what the issue meant for our assessment. Because of the known errors in the XBTs, we concluded that the ocean data showing a rapid warming in the 1970s followed by a cooling in the 1980s did not negate the overall strengthening of evidence for human influence on climate. We were well justified in calibrating our attribution statement as being 'very likely' that increasing greenhouse gas concentrations caused most of the warming since the mid twentieth century rather than 'likely'. Importantly, though, the issues with the ocean data did prevent us coming to an even stronger conclusion, such as the attribution was 'extremely likely' (meaning greater than 95 per cent probability) or 'virtually certain' (greater than 99 per cent).

These technical details about our assessment were running rapidly through my mind as Susan once again called China to speak after another series of supportive interventions from other countries who were keen on the 'very likely' statement. The representative of the world's most populous nation still did not agree. He did not regard the additional evidence of change from across the climate system – the ocean, the ice, the winds – as a big step forward from the 'likely' attribution statement of the previous IPCC assessment. He insisted that the appropriate level of likelihood hinged on the evidence from surface temperatures.

At such moments of difficulty, the chair of the meeting has a decision to make. She could, if she wished, do what her predecessor had done in Shanghai six years before when the Saudi delegation had objected to the previous report's headline attribution statement. Then the discussion had been sent into a side room for a small group of government delegates and scientists to thrash out a solution. But for Susan, sending the debate elsewhere wasn't an option. Kicking the can down the road like this would send a message to delegates that we scientists were struggling to convince governments of the correctness of our arguments. In Shanghai, the disagreement had been about the interpretation of a single word – 'substantial'. It was a disagreement open to resolution by a small group looking for an alternative wording that would fit the bill for the magnitude of human-induced warming.

Here in Paris, the disagreement was much more fundamental. It was about trust in the scientific claim of a growing strength of evidence for the effects of human activities on climate. This dispute couldn't be kicked elsewhere. We had to resolve this right now, all together.

With the clock ticking on, Susan invited another round of interventions. As she worked down a list of countries who had indicated they wanted to speak by raising their country nameplates to the vertical, Gabi, Francis, Neville and I listened with relief as delegate after delegate reiterated their support for our assessment. Our conclusions were robust they told us, strong and based on solid science. Some delegates even claimed the evidence would support a 'virtually certain' statement. They were probably making a rhetorical point rather than a serious suggestion. They knew we couldn't support such a high level of confidence because of the issues with the ocean temperature data. But they were sending a signal that they trusted our scientific judgement that the evidence for human-induced climate change had clearly grown since the last report.

Susan came back to the Chinese delegation once more. Still they refused to back down. Where was the evidence that we had made substantial progress in the science of attribution, they wanted to know? With alarm, I realized that the line of the Chinese attack was now moving on to my turf, the reason I had been asked to come here. Maybe Susan would ask me to address the delegates, a prospect both gratifying and terrifying. But instead, Susan left her seat and walked across the podium towards us.

She wanted to have an author conference off-mic. While the government delegates looked on at the silent pantomime as we scientists engaged in animated discussion, we considered the suggestion of our author-in-chief that we add a clause to our statement elaborating on the remaining caveats. We could add this, she explained, to get the statement through. I hated the idea and said as much; it would dilute the strength of a statement that was totally justified by the detailed lines of reasoning set out in the report. The others agreed. Hearing our strength of feeling, Susan accepted our position. In that case, she said, she was going to have to carry on trying to get our statement through. She walked back to her chair in the middle of the podium and looked around the room for whom to call next.

It was at this moment of greatest peril that help came in the unlikely form of an intervention from the delegate of Saudi Arabia. He wanted to propose, he said, a different way of stating the evidence on the causes of climate change. Instead of the original wording, which had the observed increase in greenhouse gas concentrations *causing* global average temperatures to rise, he suggested that a different form of words would be much better. In his alternative formulation, the rising temperatures were *due to* the increase in greenhouse gases. Neville looked across at me and nodded and I looked across at Gabi and Francis and similarly affirmed my assent. We didn't want the delegates in the hall seeing us looking too delighted but there was no doubt this suggestion was one that represented no change whatsoever in substance and was therefore one that we could readily agree to.

Susan confirmed that the authors could accept the proposal and took another round of interventions from the floor. Delegate after delegate spoke up in support, arguing that this reflected an important and substantial change that carefully responded to the difficulties some delegations had had with the original proposal from the authors.

The new formulation was an improvement, the spokesman for China accepted when Susan called on him again. But still, he wanted to know, wasn't the new IPCC report just using the same old tools as the previous one? If so, how could we argue for stronger conclusions than last time? Hadn't there been any new methodological developments since? And then I remembered the paper with Chris Huntingford, the one setting out the new methodology that took account of possible errors in variables from climate models.

'Yes, there have,' I whispered excitedly across to Francis, who like me was also very interested in the statistical basis of attribution, 'EIV!' Prompted by the flurry of activity to her left, Susan gave the microphone to Francis who, recalling our discussion in the chapter team about this new form of optimal detection, explained the intricacies of EIV to the assembled throng. Gloriously for a mathematician like myself, these statistical details were not – as they would normally be to a group of politically appointed diplomats – technical minutiae of little direct relevance or interest. Instead, as it turned out, such details provided the key that unlocked the previously unsolvable impasse created by China's initial objection to the attribution statement. It was an objection they had initially raised over an hour and a half ago. But now, at long last, their last argument fell away. An agreement was now firmly in our sights.

And yet, given their previous intransigence, if we were going to reach the promised land of a headline statement agreed in consensus by all governments present, we were going to have to allow China a small concession to smooth over their eventual acquiescence. We must, their delegate insisted, reflect the remaining difficulties in precisely attributing past climate change by including a footnote noting that 'consideration of remaining uncertainty is based on current methodologies'. An ingenious suggestion, it allowed China to claim their intervention had improved the document without adding anything that we could not support or taking anything away from our agreed headline statement. We agreed to the footnote which was typed into the document displayed on the big screen. There were no further objections and Susan finally recorded the headline statement, 'Most of the observed increase in global average temperatures since the mid-20th century is very likely due to the observed increase in anthropogenic greenhouse gas concentrations.'

I was delighted that the science had prevailed despite challenges from countries who might prefer the science to be otherwise. The new IPCC assessment when it was publicly released the following day would contain the strongest statement yet linking past global warming to human-induced emissions of greenhouse gases. As a team we had achieved what we set out to do: see all the countries of the world accept that climate change posed a real and present danger. And through my own research, my labours writing and rewriting the relevant section of the report and my coolness at a critical moment of drama in spotting the crucial pertinence of one specific scientific advance, I had been able to play my own part. I could go away feeling proud of what I had achieved.

For many of my colleagues from the chapters concerned with projections of future changes in climate, their moments of tension were yet to come. Ahead of us lay a lengthy final session that stretched long into the evening. But buoyed by the knowledge that the attribution section was now behind us, progress through the rest of the document, if slow, was steady. With no further major hold-ups, Susan gavelled down the final sentence of our Summary for Policymakers just before midnight.

Across the hall, delegates and scientists alike stood and applauded. At that moment, I couldn't readily recall the chain of events that, remarkably, had found me taking part in such a momentous event. But I felt immensely happy that I was able to share in the celebratory mood, not just with my fellow authors from the chapter on the causes of climate change, nor just with the wider team of authors who had worked across the whole report, but also with the larger IPCC family – the women and men who had worked together to construct the most authoritative, most comprehensive and most robust assessment of the science of climate change that had yet come into existence.

Susan gathered us authors together on one of the balconies of the centre, where she gave us a brief valedictory address praising us for our team's spirit and resolution, and we dispersed gleefully into the deserted Parisian streets. Gabi, Francis and I found a bar still open in the early hours and drank a toast to a successful report. What we had just witnessed felt like a significant turning point. Now we hoped that people would finally start taking the issue of climate change more seriously and their governments would commit to the sorts of reductions in emissions the gathering crisis demanded. They had just agreed, unanimously, the strongest worded statement yet blaming human activities for global warming. We had given them the ammunition to act. Now, we surely believed, they really would. It felt like an achievement well worthy of celebration.

When I arrived back at UNESCO HQ just before ten the next morning, the setting for four days of ardent debate had been transformed. In place of the government representatives who were now on their way home sat hundreds and hundreds of reporters. I took a seat in the section at the front reserved for the authors of the report and gaped at the spectacle. Every space behind me was taken, and along the edge of the room camera crews jostled for the best positions from which to film the upcoming proceedings. At the front, officials bustled in and out, fussing over the water jugs and microphones for the panel of dignitaries who had come here to help launch the report. And behind me, journalists were laughing and smiling when they saw that I, like many of my colleagues around me, was taking pictures of them, the people waiting to hear from us, the authors of this landmark report.

I had gone to bed very late and struggled to get to sleep after the dramas of the day before, but this atmosphere of intense anticipation felt immensely energizing. I had no idea it was going to be like this. Closeted inside this concrete bunker of a building for the best part of a week, focused entirely on one aim, that of supporting Susan and the rest of the team in securing the report, I had lost touch with whatever reality lay outside. Now, emerging into the glare of the world's media, I could see that our report was very big news indeed.

Susan kicked off the press conference with a short presentation. She looked tired, her voice a strained monotone as she summarized the main messages of the report. Captaining its production had been a multi-year marathon that had engaged every sinew of her formidable intellect and willpower. Her own finishing line had been reached just a few hours earlier with the approval of the final sentence of the Summary for Policymakers. It looked like she was dreading the prospect of addressing the media.

In contrast, refreshed, smartly suited and with sound bites at the ready, Achim Steiner, executive director of the United Nations Environment Programme, was clearly in his element addressing such a large crowd of reporters. A media-savvy speaker, he set a triumphant tone. 'Friday, 2 February, 2007 may go down in history', he said, 'as the day when the question mark was removed from the question of whether climate change has anything to do with human activities.'[12] The UN supremo's eye-catching statement had highlighted the central importance of what our attribution science had achieved in pinning the blame for global warming squarely on human not natural causes. It was a historic moment.

Then it was on to questions from the floor. Susan fielded them in the same measured and downbeat manner that she had given to her presentation. 'Was it even worse than we thought?' one journalist asked. 'How did she feel?' another wanted to know. 'What was her message to governments?'

Painstakingly, our author-in-chief stuck to the carefully crafted scientific messages set out in our report. Although at times it felt strained not rising to the bait of journalists wanting emotive language, I knew she felt strongly that this was the best way. Our job was to explain climate change was real and posed a huge threat, she thought, not tell governments exactly what they should do or opine on how we felt personally. Our report was based on established evidence not subjective opinion.

Opening formalities completed, journalists hurried over towards the scientists gathered in our corner of the room. A huddle of reporters gathered round Phil Jones to hear more about the headline statement that 'warming is unequivocal'. As coordinating lead authors, Gabi and Francis took the lion's share of media interest in the attribution statements.

Who I talked to was all a bit hit and miss. The IPCC had not yet understood the full significance of cultivating the media and of putting resources towards disseminating its results to the public. They had sent only one press officer to Paris and she was completely overwhelmed with the scrum of journalists looking for authors to talk to. At one point, not realizing I had been slated to appear live on TV rolling news, she found me taking a breather in a corner and rushed me across the centre and out into the courtyard where the television crews were stationed. Hurriedly greeted by a relieved reporter from the BBC and rapidly clipping my microphone on to my shirt, I took my first question live on air seconds later. The main BBC ten o'clock television news that evening was presented by Fiona Bruce live from the same spot. There was only one lead story that night: the news we had made in Paris.

The next day, all the newspapers led with our report. 'It's worse than we thought', said the *Guardian* on its front page.[13] Most highlighted the new attribution statement. 'Human activity is very likely to blame', said the *International Herald Tribune*, accurately capturing the key part of the wording we'd worked so hard to defend.[14] The *Financial Times*, presumably eliding the unequivocal warming sentence with the attribution one, went with 'World's scientists certain that humans cause global warming'.[15] If some newspapers made small mistakes like this, I wasn't too concerned. They had got the main point across that scientists agreed that climate change was real and it was largely down to human activities. The publicizing of this message was already long overdue.

This extraordinary level of worldwide attention felt like vindication for all the hard work we'd done, marshalling the evidence and convincing governments of our conclusions. The carefully chosen words of our report that had been unanimously agreed here in Paris would surely guide the upcoming climate negotiations due to culminate in three years' time with the signing of a global agreement in Copenhagen. I was hugely optimistic that governments would act. It felt like we had reached a turning point in the battle to halt climate change.

Already, many politicians weren't shy to make the most of the opportunity presented by the IPCC's new report. British politician, David Miliband, Secretary of State for Environment, Food and Rural

Affairs, called for international commitment to take action. The report, he said, 'represents the most authoritative picture to date, showing that the debate over the science is well and truly over.' It was, he said 'another nail in the coffin of the climate change deniers.'[16]

The same day as the release, the Department for Environment, Food and Rural Affairs and the Department for Education and Skills jointly made an announcement: *An Inconvenient Truth*, a movie setting out the science of climate change and documenting former US vice president Al Gore's mission to inspire action to prevent further damage, would be sent to every secondary school in England. According to Miliband, 'Everyone can play a part along with government and business in making a positive contribution in helping to prevent climate change.'[17] The British government's idea was that distributing a copy to every school would stimulate children into discussing global warming in class. Their future was at stake, after all. It was a move that would land me in court.

7

Court of opinion

On 27 September 2007, I hurried along the Strand in London towards the imposing Victorian Gothic edifice of the Royal Courts of Justice. The government had been accused of breaking the law in distributing the highly publicized climate change film *An Inconvenient Truth* to schools, and I was the expert witness they had called in their defence. The hearing of the case, which relied heavily on my evidence, was about to begin.[1]

Nevertheless, before entering through the iron gates and under the elaborately carved entrance porch, I hesitated on the pavement outside where on television news bulletins I had seen triumphant litigants relishing their victory. I couldn't help worrying about what was at stake. If my evidence didn't carry the day, in a few weeks' time this scene could hold a crowd of eager journalists gathering round the vindicated claimant as he berated the establishment for brainwashing his children with illegal propaganda. What a triumph that would be for him. What a disaster that would be for me.

Not because I would have failed to prevent damage to the credibility of the British government. That wasn't my main concern. What mattered to me more was the credibility of the scientific message on which I had worked so hard for so many years. If we lost, TV and radio bulletins would announce that the highest court in the land had ruled that the threat of climate change had been exaggerated to serve dubious political ends. I would have failed to prevent children missing the real news: that climate change was a serious problem that threatened their futures and needed to be addressed urgently. The damage would be untold. Beyond the turreted stone walls rising up before me, a distinguished judge would soon start considering the testimony that I had provided. So much of the future of climate action, it seemed, depended on me.

Inside the echoing interior, the courtroom where my case was going to be heard was off an upstairs corridor halfway along on the right. Near its thick double doors stood Martin Chamberlain, the government's barrister, dressed for action in wig and gown and waiting to give me some last-minute instructions. The court would be guided by my two detailed witness statements, he told me. I wouldn't be called for further oral evidence today but if there was anything that I wanted to point out to him during proceedings I could send up a note.

The defendant's grilled compartment stood against the left-hand wall of the oak-panelled courtroom, unoccupied. It had a smell reminiscent of my old school's classrooms, of furniture polish, dust and stress. Representatives of Her Majesty's Government were not expected to answer directly to the charge from a position in the dock. Instead, it was Chamberlain, the barrister acting on their behalf, who would set out their defence. But the absence of ministers brought into sharp focus that it was not really them on trial, even though they were the ones who had decided to distribute the film to schools. The film was about climate science and what it meant. It was that science that stood accused here today.

The movie *An Inconvenient Truth* provides a popular account of the basis of global warming. Released in 2006, it is presented by Al Gore, former vice president and lead American negotiator of the 1997 Kyoto agreement to limit greenhouse gas emissions. The film was a huge hit around the world, surprising perhaps given that it is a documentary film about a topic that had at that point received little widespread attention. Based around a sophisticated PowerPoint presentation that Gore had given over a thousand times across America, it summarizes the scientific evidence on global warming and tells the story of his mission to promote greater awareness about the climate crisis. It was widely seen in cinemas in the US and across the world, and won many accolades including Best Documentary Feature at the 2006 Academy Awards. It was praised by critics and climate scientists alike for its powerful and informative presentation.[2]

It also received a lot of attention in the UK. But in March 2007, the British public were on the receiving end of a disastrous piece of television, a polemical documentary broadcast on Channel 4 to an audience of 2.5 million that claimed global warming was a lie. Entitled *The Great Global Warming Swindle*, it featured many of the climate deniers I had met in Moscow, including Richard Lindzen, Paul Reiter and Piers Corbyn, as well as Patrick J. Michaels and former British Chancellor of the Exchequer Nigel Lawson.[3] Nothing that they said was consistent with the well-established scientific consensus. Given that many pupils and their parents might have been influenced by this farrago of nonsense, I thought ministers were right in sending *An Inconvenient Truth* to schools. But that motivation to inform young people properly about the nature and seriousness of climate change had led to the claim that the government had broken the law.

In recent weeks, since I had been asked to provide expert testimony on the veracity of Gore's film, it had become a professional obsession. I had spent many hours assessing its scientific accuracy and in two detailed witness statements to the court I had set out my views as to why the film was neither error-strewn nor biased as the opposition had made out. Now, finally, I was about to find out what the judge allocated to the case made of it all. My assessment of the scientific realities was about to be confronted by the legal realities of a British courtroom.

The door at the front of the court opened, we all rose and a clerk entered followed by the judge, Mr Justice Burton. Once he had seated himself, the rest of us were allowed to sit and listen to his opening remarks. He addressed us in a relaxed manner, like a gracious host at a niche form of entertainment. Although a previous judge had indicated that one day would suffice for an oral hearing, our judge did not agree with his learned colleague. More than one day would be needed to do justice to the arguments involved. He would give the case as much time as necessary.

Now he came to the matter in hand. He hadn't seen the film. A social engagement the night before had prevented him watching it on the DVD he had been provided with. Instead he was going to hear the views of the claimant's counsel first. He would take a look at the movie tonight. I was no expert in legal etiquette but the judge had not only failed to take the trouble to watch the central feature of this legal drama in advance, he was going to let the prosecution lay out their detailed case against it before he did so. This felt like a deeply inauspicious start.

*

Several months previously, long before I knew anything about a possible legal case, I had gone to see *An Inconvenient Truth* with colleagues from work at a local cinema. This was the first time that climate science had broken into the mainstream in this way. A group of us wanted to go together so we could discuss what we thought about it later in the pub.

I enjoyed the movie's accessible presentation of a wealth of scientific information about the causes and effects of climate change. Gore talks about being inspired by his college professor, Roger Revelle, the man who had the idea to measure the amount of carbon dioxide in the atmosphere and who hired Charles David Keeling to set up measurements on the Mauna Loa volcano. Gore describes his political journey, organizing the first hearings in the 1970s on global warming in Congress and coping with the blow of losing the 2000 presidential election. And the American politician ends with a call for action. His country should lead the world in ending its dependency on fossil fuels. This is a moral issue, he says. It is time to rise again to procure a better future for everyone.[4]

Discussing the film over a beer with my fellow climate scientists, we all liked its accurate presentation of climate science and welcomed the way Gore eloquently sets out some of the devastating impacts of climate change. I particularly liked the way the film carries an emotional punch. It seemed like a punch that would hit home with the people who mattered most – youngsters in school who should know what awaited them if action was not taken soon to reduce our emissions.

But it was apparently not to the taste of Stuart Dimmock, a parent and school governor from Dover in Kent. A member of a small political organization called the New Party which promoted climate change denial, Dimmock had engaged a legal team to submit a written application challenging the government's actions in sending the film out to schools.[5] With the heavy legal costs involved, he needed strong financial backing.

Dimmock was supported by a powerful network of business interests linked to fossil fuels and mining.[6] The New Party received almost all its funding from a Scottish quarrying company whose owner, Robert Durward, was also the chairman of the party. Durward had co-founded a lobby group called the Scientific Alliance which in December 2004 teamed up with American lobby group, the George C. Marshall Institute, a long-time promoter of climate change denial, to produce a briefing paper attacking climate science. The Scientific Alliance had brought climate change denial lobbying to the UK.[7]

In the US, there was a long history of well-funded lobby groups attacking the consensus findings of science. The George C. Marshall Institute founded in 1984 was one of the oldest, originally set up to support Ronald Reagan's 'Star Wars' programme of missile defence by attacking scientists such as Carl Sagan.[8] By the early 1990s it had moved on to environmental matters, first attacking the science of ozone depletion and then the science of global warming. It was influential in lobbying the incoming Bush Senior administration in early 1989 not to take climate change seriously. Its founder Frederick Seitz was one of the principal 'climate sceptics' who incorrectly and maliciously accused Ben Santer of fraud in 1996, and many well-known climate deniers including Patrick J. Michaels and Richard Lindzen were involved in its activities. When in 2001, the Global Climate Coalition disbanded after George W. Bush rejected joining the Kyoto Protocol, it was the George C. Marshall Institute who led the continued resistance to climate action in the US. It supported Congressional briefings involving Senator James Inhofe, Chair of the US Senate Committee on Environment and Public Works, who called man-made global warming 'the greatest hoax perpetrated on the American people'. And it received substantial funding from Exxon Mobil.[9]

Its British equivalent, the Scientific Alliance, helped coordinate the activities of climate change denial in the UK. It organized conferences involving climate deniers, helped ensure prominence for the views of climate deniers in newspapers like the *Daily Mail* and advised Channel 4 on *The Great Global Warming Swindle*.[10] With millions of readers and viewers on the receiving end of such climate disinformation, the Scientific Alliance seemed to be having some success in damping down engagement by the British public in the issue of climate change. *An Inconvenient Truth* threatened to change all that.

On its distribution to schools, Alan Johnson, Secretary of State for Education, had said that children were the key to changing society's long-term attitudes to the environment.[11] Watching and then discussing the film, he said, would help them influence the lifestyle and behaviour of their families. Supported by the Scientific Alliance, Stuart Dimmock sought to prove that the government had broken the law by attempting to indoctrinate pupils.

Sections 406 and 407 of the 1996 Education Act stipulate that 'The local education authority, governing body and head teacher shall forbid the promotion of partisan political views in the teaching of any subject in the school' and 'shall take such steps as are reasonably practicable to secure that pupils are offered a balanced presentation of opposing views'.[12] In sending *An Inconvenient Truth* to schools, the Department for Education and Skills had also provided teachers with an online resource including guidance notes which put the film into the wider context of scientific understanding. In their written submissions, Dimmock's legal team argued that showing the film in schools promoted partisan views and therefore broke the law according to the Education Act. They claimed school pupils were not receiving a balanced presentation of evidence for and against climate change as a serious issue.

The claim for judicial review of the government's original decision to distribute the film to schools took some time to come to court. A previous judge had assessed the case on the papers and had refused the application to take it further. But after appeal, it was eventually agreed that a 'rolled-up' hearing should be held at which permission to proceed would be followed immediately by consideration of the substance of the matter.

This is when I got involved. Knowing my involvement in the IPCC and wanting a scientist from the Met Office Hadley Centre, the institute that had helped develop the online guidance notes for teachers, civil servants had approached me to see if I would be prepared to be instructed by the Secretary of State for Education and Skills to act as an expert on climate change. Keen to do my civic duty and intrigued as to what this new experience might involve, I said yes. At that point, I had little idea what such a role would eventually entail.

In June 2007, with some trepidation, I travelled up to London for my first meeting with the government's appointed counsel, Martin Chamberlain, at his chambers in the Temple. Feeling very much on foreign ground, sitting in front of an imposing desk in a plush office at the heart of my country's legal establishment, I listened attentively to what the barrister had to tell me. He spelt out that testifying in court was a heavy responsibility. I was strongly advised to take particular note of the rules regarding experts in a court of law, he said, rules that laid out my duty in assisting the judge by offering an independent opinion based on a sound foundation of facts.[13] Above all, I must remember that my overriding duty was to the court not to the party from whom my instructions had been received. I could forget that at my peril.

With this heavy introduction out of the way, the barrister softened a bit. He was keen to hear what I really thought of the movie, deep down. Did I like it? Did I think, with the benefit of my extensive experience in climate science, that the film had got things mainly right or were there some aspects of a complex subject that it had actually got wrong?

My summary was that its presentation of the causes and likely effects of climate change was broadly accurate. Where there were a few shortcuts, inevitable in any popular presentation of complex science, the relevant details had been set out in the accompanying guidance notes sent out by the Department for Education for teachers to refer to when they discussed the film in schools. Having seen the work of colleagues who had helped the civil servants draft those notes, I knew they were pretty comprehensive. I said to Martin Chamberlain what I'd said to my colleagues when we'd first seen it at the pictures: I liked it not just for its accurate depiction of the science but for its wider emotional appeal.

My basic standpoint established, the meeting then proceeded to discussing how I would go about writing my witness statement. I needed to establish the main causes of climate change as widely accepted by the scientific community. I should also consider the arguments put forward in the claimant's skeleton argument as to why the Education Act had been breached. Furthermore, I needed to comment on a witness statement provided by Robert Merlin Carter, a retired professor affiliated to James Cook University, Queensland. Carter was a geologist and outspoken critic of climate scientists who dismissed the ever-growing wealth of evidence showing the seriousness of human-induced climate change. He was one of the world's most vocal and best-known climate deniers.

In preparing my text, the barrister advised, I should rely heavily on the findings of the Intergovernmental Panel on Climate Change (IPCC). The recently published report that I had seen approved in Paris was the most comprehensive, well-reviewed and widely agreed assessment of the state of the science available. Having helped write it I should know its contents as well as anyone. And as an IPCC lead author I had the necessary authority to explain its main messages to the court.

This was my task then: to take Al Gore's movie on the one hand, and the hundreds of pages of the latest IPCC report on the other, and advise the court as to whether or not *An Inconvenient Truth* faithfully represented the causes and likely effects of climate change. In doing so, I was going to have to address my opposite number's claim that aspects of the film were 'substantially and materially at odds with generally accepted aspects of current international scientific opinion on climate change'.[14]

The Australian climate denier had set out three areas where he claimed the film had got it badly wrong. These were the evidence of atmospheric change from ice cores, the melting of the polar ice caps, and the linking of extreme weather events and disasters to global warming. I had to deal with each one in turn, explaining what the scientific understanding really was.

The ice cores showed a rich record of carbon dioxide concentrations varying in step with global temperatures over the last several hundred thousand years. The science was well established. As the ice ages came and went, plunging the earth in and out of a succession of deep freezes, so the carbon dioxide in the air plunged and soared, from fewer than 200 parts per million (ppm) in the glacial eras to 280 ppm in the interglacials. Then, in the industrial era of more recent times, the carbon dioxide had shot up towards 400 ppm.

An Inconvenient Truth illustrates this relationship with an animated chart, showing how the carbon dioxide and temperature levels tracked each other over the last 650,000 years until, right at the end, in the industrial era, carbon dioxide had risen so high that Gore has to step on a lifting platform in order to trace the line as it races up towards the ceiling of the lecture hall. The point that carbon dioxide and temperatures were intimately related had been vividly made. And the implications for global temperatures had been hammered home, if carbon dioxide concentrations rose to ever greater heights above what had previously been seen in the ice core records.

In his witness statement, Carter claimed that far from increases in carbon dioxide concentrations causing the global warming seen in recent decades, cause and effect was the other way around, that warming from natural variations in climate then caused the increases in carbon dioxide. All reputable scientists know that this claim is incorrect. But the reason why is relatively complex to explain. In my witness statement, I had to account for the intertwined relationship between the two, not just in recent decades but over millions of years.[15]

The recent geological past is marked by the cycling – on a roughly hundred thousand-year timescale – of the earth in and out of ice ages. It begins with small fluctuations in the planet's orbit around the sun, which change the distribution of solar insolation arriving at the surface of the earth. When northern latitudes receive less sunlight in summer, ice sheets expand, brightening the surface, which then reflects more solar radiation back to space. Sea levels fall, exposing new land on which vegetation can grow and absorb carbon dioxide. The oceans cool and as a result they too absorb more carbon dioxide from the air. Greenhouse warming is reduced, temperatures drop some more and yet more carbon dioxide is absorbed by vegetation on land and the colder oceans. Greenhouse warming is reduced even more.

It's an example of a climate feedback, a ratcheting-up process that has made the earth prone to large swings in climate on geological timescales. Over the last million years, this feedback process has meant lurching from the harsh conditions of the glacial periods with their massive ice sheets covering huge areas, to the much more temperate and habitable times of the interglacials, as the earth's orbital parameters subtly shift. Over the last 10,000 years of temperate interglacial climate, much of the story of human civilization has taken place. Thanks to favourable orbital parameters and a relatively constant sun, such a stable habitable climate would have carried on for many thousands of years more. But now human emissions of greenhouse gases have begun adding to the greenhouse effect and warming up the climate. As Al Gore explains, these added greenhouse gases are now causing rapid warming. It is not the case, as Robert Merlin Carter claimed in his witness statement, that Gore's presentation is 'substantially and materially at odds' with scientific understanding of this aspect of climate change.

The take-home message from this story of climatic upheaval is that the earth's climate is highly sensitive to small perturbations. Humanity has benefited from an extended period of stable global temperatures. Since the start of the industrial era, that favourable climate has been put at peril. Global temperatures have responded to increasing greenhouse gases in the atmosphere by warming up. In the last few centuries, the rising carbon dioxide concentrations from human activities have significantly perturbed what would otherwise have been a fairly constant climate.

Everything I wrote in my witness statement was well-accepted scientific truth. I knew the explanation for the relationship between carbon dioxide and temperatures over the last 650,000 years was not that straightforward. Gore himself had said so in the movie. But the guidance notes sent to schools provided all the context teachers needed to avoid pupils being misled. If mischief makers or skilful barristers could twist a relatively complex truth to obscure the simple reality of human-induced global warming, my job was to set out the scientific facts as clearly as I could.

Having dealt with the first of Carter's claimed errors, I moved on to his second claim, that Gore had greatly exaggerated the impact of future melting of Greenland and Antarctic ice. Whereas the American had stated that human actions would lead to 20 feet or more of sea level rise, the Australian maintained that the real figure over the next hundred years was less than 2 inches. The depiction in *An Inconvenient Truth* of cities being submerged under water, Carter alleged, was pure fantasy.

Gore, I explained to the court in my written witness statement, had faithfully represented the impact on sea level of the melting of the Greenland and the West Antarctic Ice Sheets. By raising atmospheric temperatures to a point where the large ice sheets could not survive, their demise would indeed raise sea level by many feet (or metres). This was a clear and present danger of continued emissions: that we could soon reach a point beyond which some of the earth's largest ice sheets could not be saved.[16]

Gore, it was true, had glossed over the long timescales involved in melting all those millions of cubic metres of ice. To completely submerge large parts of San Francisco Bay, Shanghai and Manhattan, as depicted in the film, would take many centuries. For gigantic bodies of ice to melt takes time, even if their very existence is, as Gore rightly implies, imperilled by human actions over just the next few decades.

But if Carter had spotted a gap in the scientific detail of the film, perhaps not surprising in a ninety-minute movie designed for widespread popular appeal, the accompanying notes sent to schools explained the long timescales involved. Pupils *were* being provided with material which, taken together, was entirely consistent with the IPCC's latest assessment. In contrast, Carter's claim that sea levels would rise by less than 2 inches over the next century was total nonsense.

Carter's third accusation was that Gore had attributed 'some twenty current or projected climatic events or disasters' to human-induced global warming, which was in every one of these instances 'at best doubtful and at worst (in most cases) known to be scientifically incorrect'. These disasters included the melting of the polar ice caps, increased severity of storms and catastrophic loss of coral reefs. These examples were fairly easy to deal with. All three were climate impacts that had been demonstrably linked with human-induced greenhouse gas emissions.[17] But there were others in Carter's lengthy list of 'events or disasters' he claimed were nothing to do with climate change that were trickier for me to respond to.

An Inconvenient Truth features distressing images of people pleading for help in the aftermath of Hurricane Katrina and the drying out of Lake Chad, once the sixth largest lake in the world. These are both graphic examples of the sorts of disastrous storms and droughts people are likely to face more and more as climate change progresses. But at the time when I was writing my witness statement, neither event had been robustly attributed to human-induced emissions, although evidence has since grown that global heating is making hurricanes like Katrina more intense.[18] Viewing the film again, it was clear that Gore did not claim the disasters could unambiguously be blamed on climate change. Rather he used the footage as justifiable illustrations of the terrible devastation that further global warming would bring.

Nevertheless, I had a duty to the court to respond to Carter's statement in full even if Carter was tempting the enquiry on to dubious ground by equating Gore's illustrations of climate change with attributable events. Whether or not I thought this was a spurious line of enquiry, I had to set out to what extent these events could, according to current scientific understanding, be attributed to human activities on climate. I carefully explained that 'evidence remains insufficient to establish a clear attribution for some particular examples shown in the film, including the drying out of Lake Chad'.[19] I also provided information about the expected impacts of climate change which the film had chosen to highlight and these included the effects on making hurricanes more intense. As I wrote, 'Gore's use of hurricanes as an example of the effects of global warming is therefore justified.'[20]

Writing all this down in a way that was both scientifically correct and in the appropriate legal format took a lot of time and effort. When it was finally submitted, I felt a sense of relief. I'd answered Carter's erroneous criticisms as best I could and demonstrated why Gore's movie was faithful to our scientific understanding, especially when supported in classrooms by the government's guidance notes which colleagues at the Met Office Hadley Centre had helped put together.

Then, with just a few days to go before the oral hearing, the claimant's legal team played a dubious tactic. They provided a further witness statement from Carter, this time stretching to thirty-one pages and involving seventy cited academic references. It now accused *An Inconvenient Truth* of making twenty itemized scientific errors, some of which had been raised in his previous statement but many of which were new. Carter's latest effort, which drew on many of the erroneous arguments used by climate deniers over the years to try to discredit climate science, was a litany of mistaken allegations. This new riposte could not be left unchallenged. It was down to me to respond.

With little time left before our scheduled date in court, I was given three days to turn my response around. It would have been challenging enough had I been based at my office with no other commitments to worry about. But I had already travelled to Edinburgh to attend the annual conference of the Royal Meteorological Society. I had a talk to give, hadn't brought a laptop with me and had no easy access to the internet. I was forced to make the best of an unpromising situation.

I borrowed a fellow delegate's laptop, abandoned all ideas of attending any of the talks apart from my own, and got to work. At the end of the second day, exhausted by wading through hundreds of pages of detailed information about the twenty topics identified by Carter, I took time out and went to the conference party.

It was a Scottish ceilidh with an energetic band and a caller inviting the crowd to join in dances like Strip the Willow and the Gay Gordons. The music was loud, the dancing was sweaty and the mood was exhilarating. Almost everyone joined in, taking their partners by the hand and doing the do-si-do, young and old, student and professor, postdoc and departmental head. After hours spent engaging with the mind of a climate denier, kicking back like this with my community of fellow weather and climate researchers felt like balm for the soul. For an hour or two, I was able to forget about the pressure of my imminent deadline and the frustrations of having to laboriously rebut cleverly constructed but spurious arguments.

Walking back across campus under a starry sky to my room in student halls, I felt a sudden surge of optimism. The state of knowledge on climate had been advanced by the sort of open-hearted bright-eyed people I'd spent time with that night. Unlike the so-called climate sceptics, these were people trained to be truly sceptical, always questioning, not prepared to take statements as read unless they could stand up to rigorous scientific scrutiny. Surely it was their way, not that of climate change denial, that would win out in the end.

And then, walking down the gloomy corridor to my temporary digs, I felt a sudden countervailing pang of anxiety. Having felt so happy in their fellowship now brought home to me the responsibility I had to faithfully represent them, to properly describe their findings without undue caveats or understatement but also without oversimplification or exaggeration. It felt a daunting task in what had to remain a relatively brief legal document. And the next day I would have to finish it, else miss the boat in responding to Carter's eleventh-hour submission.

Back at my temporary desk the following morning, I realized the only way I could respond fully to Carter's claim of twenty scientific errors in *An Inconvenient Truth* was to compile a table, listing a rebuttal for each of the alleged errors. In numerical order, the claims in the film that were supposedly mistaken were that: (1) the earth's temperature is increasing dangerously; (2) before the industrial era, atmospheric carbon dioxide did not exceed 300 ppm; (3) increasing carbon dioxide will cause dangerous warming; (4) carbon dioxide is the most important human-induced greenhouse gas; (5) carbon dioxide is a pollutant; (6) computer climate models can provide accurate predictions of future climate change; (7) the European heatwave of 2003 was caused by human-induced global warming; (8) Hurricane Katrina was caused by human-induced global warming; (9) hurricanes will become more powerful because of global warming; (10) higher insurance costs for weather disasters are a reflection of global warming; (11) sea level rise of up to 20 feet (7 metres) will be caused by melting of West Antarctica and Greenland; (12) low-lying inhabited Pacific atolls are being inundated because of global warming; (13) coral reef bleaching and species loss is being exacerbated by global warming; (14) Mount Kilimanjaro and other mountain glaciers are being melted by human-induced global warming; (15) polar bears are threatened by global warming; (16) Lake Chad dried out because of global warming; (17) drought in the Sahara has been caused by global warming; (18) the ocean conveyor belt in the North Atlantic will be deleteriously affected by global warming; (19) malaria and other tropical diseases are being exacerbated by global warming; and (20) that the film ignored the dominant role of natural variations in water vapour and solar radiation in driving recent climate change.[21]

In compiling my table, I went through each in turn, drawing on the scientific evidence set out in the latest IPCC report. For sixteen of the twenty alleged errors, it was clear that Carter was demonstrably wrong. And for the remaining four, Carter was drawing the court towards a misleading interpretation of how the film would be viewed in schools in conjunction with the guidance notes.

For Hurricane Katrina, I had already explained in my first witness statement that Gore's use of hurricanes as an example of the effects of global warming was entirely justified. As for the drying out of Lake Chad and Saharan drought, these were being used as illustrations of the effects expected in many subtropical regions as rainfall declined with climate change.[22] And while the melting of the Kilimanjaro snows was not the best illustration, given that the causes are complex and related to local factors, the many other glaciers shown in the film – in Alaska, the Himalayas, the European Alps and South America – were well chosen to illustrate the widespread melting of glaciers due to climate change. Overall, Gore had based his film squarely on compelling scientific evidence of the deleterious effects of ongoing emissions of greenhouse gases. And while there were some aspects of the film where further context would be helpful in the classroom, these were provided by the guidance notes supplied to teachers.

I finalized the drafting of my second witness statement and completed my table of responses to the twenty 'errors' in time to take the long train journey home. At the conference, I had attended only my own talk and the social ceilidh event. But I *had* been able to deliver my evidence in time to be considered by the court.

Taken together, my two witness statements and my table of responses to Carter's alleged errors provided a thorough rebuttal of the claimant's proposition that showing *An Inconvenient Truth* in schools broke the law by indoctrinating pupils. I had considered both the scientific merits of the film and how it would be used in schools when viewed in conjunction with the guidance notes. I summed up my second witness statement thus: 'I remain of the view that it could not be sensibly said that pupils who see the film presented in accordance with the guidance would receive scientifically misleading information.'

*

Three weeks later, I sat in the public gallery of a courtroom at the High Court of Justice and listened to Judge Mr Justice Burton invite the claimant's barrister Paul Downes to take as much time as he needed to argue the position of his client – parent and school governor Stuart Dimmock. And what a masterful performance Downes provided. In vivid terms, with a well-orchestrated appeal to emotion, he set out his case as to why the government had broken the law. Strikingly, he based his submission on a colourfully annotated transcript of the film.

Twenty per cent of the movie, Downes told us, was sentimental mush. Thirty per cent was pure politics. And of the fifty per cent that was scientific, he argued that most of it was either false or grossly exaggerated. He had helpfully marked up the different sections of the transcript in different colours. Thanks to his delineations it would be easy for his honour to tell the different sections apart when he watched the film that night. Having carefully explained which parts were which, he handed the pages to a clerk who handed it up to a grateful-looking judge. It seemed outrageous that the cause célèbre at the centre of this drama would be viewed for the first time in such a loaded fashion.[23]

For the rest of the day, Downes continued to set out his stall for the prosecution. Despite my annoyance with his case, I found it fascinating how he presented it, his dramatic delivery and theatrical timing, and also its legal substance and the interjections by the judge. I hadn't expected to find the law as interesting as this. I was a scientist after all, not a lawyer. But like a compelling piece of scientific research where you find yourself treading fascinating new ground in search of an answer to a question of vital interest, here legally trained minds were researching the answer to a crucial question of litigation that also needed answering: what sort of material presented in a classroom would constitute illegal promotion of partisan views?

Sections 406 and 407 of the Education Act of 1996, I learnt, had never before been tested in a court of law.[24] These Sections of the 1996 Act were actually identically worded incorporations of the 1986 Education Act which had been introduced during the Thatcher government to deal with a perceived issue of indoctrination of state school pupils by left-wing teachers.[25] The legislation was drafted to deal with concerns that such seemingly unobjectionable titles as 'anti-racism', 'anti-sexism' and 'peace studies' could be used as cover for tendentious political content.[26] Now, twenty years on, this law was being tested for the first time in the High Court.

Downes argued that the showing of any partisan film in school – as he claimed An Inconvenient Truth to be – was against the law as set out in the relevant Sections of the Act. Martin Chamberlain, in his written submission on behalf of the government, argued against. The judge seemed more sympathetic to Chamberlain's line of reasoning.[27] A history teacher could show a Nazi propaganda film in class, for example, the judge proposed, not to promote Nazi sentiment but to facilitate pupils' learning about the past promulgation of views with which they may well vehemently disagree. The crucial ingredient in teaching was the context in which such a film is shown. In this case, that context concerned the guidance notes and their relationship to the film itself.

Downes was adamant that the showing of *An Inconvenient Truth* promoted a partisan view by presenting an apocalyptic vision that did not represent the scientific consensus. My evidence contradicted that opinion by demonstrating that the movie portrayed the mainstream consensus view, albeit in a dramatic and emotionally appealing way. In fact, whether the film was partisan or not was not relevant to the question at hand. After all, we had already established that showing a film of Nazi propaganda in school didn't, by itself, constitute indoctrination of pupils. What was relevant to the question, however, was whether the guidance material sent to schools alongside the film provided the appropriate balance. Only if it had, so the judge reasoned, would the law not have been breached.

It was at this point of the proceedings, just as I was finding myself pleasantly absorbed in the legalistic niceties of the case before the court, that the matter in hand came back, once again, to the scientific robustness of the film and the accompanying guidance notes. Only if together they reflected the mainstream view, the judge opined, including the guidance notes pointing out any areas where the movie had strayed from that mainstream view, would its showing in schools be in accordance with the law. Unluckily, this meant turning our attention once again to considering Robert Carter's expert testimony that the film 'goes well beyond the mainstream of scientific opinion' and furthermore had made twenty itemized scientific errors. This meant some of the country's finest legal minds trying to make sense of the abstruse technicalities of an IPCC report.

What I and my scientific colleagues referred to as the 'extremes table' is contained in the Summary for Policymakers of the Fourth Assessment Report, the document we had agreed word by word at UNESCO HQ in January.[28] It is a collation of assessments of the extent to which extreme weather events had changed up to then, had been affected by human activities, and would likely change in future assuming continued emissions of greenhouse gases. It is a complex compendium of information which had taken many hours of painstaking work. In the table, there are seven rows for seven different types of extreme weather events – including heatwaves, droughts and tropical cyclones – and three columns for the observed changes, the human-induced changes and the expected future changes. For each of the twenty-one elements in the table, a likelihood is assigned as to whether a change is occurring. According to this extremes table, heatwaves are 'very likely' to increase in future, but the extent to which human activity has already increased the intensity of tropical cyclones is assessed to be 'more likely than not'.

Now, in the High Court of Justice, two barristers and a judge tried to interpret the IPCC's use of language. Martin Chamberlain explained the specifics of IPCC's use of terms, that 'very likely' was defined to mean that there was a greater than 90 per cent chance of the statement being correct and 'more likely than not' meant that there was a greater than 50 per cent chance. But for the claimant's side, carefully calibrated scientific uncertainty was grist to their mill that the scientists weren't sure about anything. To them, such language of likelihoods where nothing was 100 per cent certain cast doubt on all the IPCC's findings on climate change. And confusion in the courtroom, as lawyers tried to come to a common understanding of what us scientists had spent three years working to formulate, could only help their case.

While I would have liked to have stood up and explained it to everyone, it was forbidden to do so from my position in the public gallery. The best I could do was pass up some notes to Martin Chamberlain when particularly erroneous errors of interpretation were made by the opposite side. But correcting such misinterpretations didn't seem to make much of a dent in the claimant's barrister's force of presentation. Seeing me handing a piece of paper forward, he interrupted his peroration to mock his opposite number for having to take assistance from an unnamed shadowy figure in the public gallery. What if scientists couldn't be confident, he summed up his morning's work, that extreme weather events were actually changing because of human-induced emissions? How much trust should we really place in the movie's emotional depiction of climate change as a sequence of devastating heatwaves and destructive storms? How could we be absolutely sure that schoolchildren weren't being misled?

After lunch, the discussions became further absorbed in the detailed minutiae of the IPCC assessment and how well they were reflected in the guidance notes. Faced with an increasingly technical discussion, the judge's interest appeared to wane. Then the claimant's barrister came to number 15 on Carter's itemized list of alleged errors in *An Inconvenient Truth*, the one about polar bears. Now the judge *was* engaged.

The movie features an animated cartoon of this iconic animal desperately swimming through open water looking for ice. In the commentary, Gore says 'A new scientific study shows that for the first time they're finding polar bears that have actually drowned swimming long distances, up to 60 miles, to find the ice. And they didn't find that before.'[29] The claimant's barrister disputed the truth of this, citing Carter's witness statement that stated, 'there is no factual evidence that warrants alarm about polar bears because of climatic warming'.[30]

The judge was all ears. 'Had polar bears drowned as Gore had claimed because they couldn't find ice?' he asked. To him, this was not a matter of confusion between meanings of likelihood, or a technical disagreement about whether the film or guidance note had correctly reflected a nuance of the IPCC's assessment. This was a manifest contradiction between two polar opposites: the Arctic bears cited by Gore *were* or *were not* threatened by global warming. What, he wanted to know, did science have to say on the subject?

As the government's expert witness, the issue of polar bears had been a difficult one for me to respond to. I was not an expert in their ecology, and the scientific literature I could find was not extensive. In the movie, Al Gore refers to a scientific study that described observations of mortality associated with extended open-water swimming by polar bears in the Barents Sea.[31] It documented aerial surveys of polar bears from 1987 to 2004. The most recent year was the first time the researchers had seen polar bear carcasses, four of them, floating in open water over 160 kilometres from the pack ice edge. They were presumed drowned. The judge took a great interest in the fate of these four bears, circling the ocean in their fruitless search for ice. It was clearly a sad situation. But where was the evidence, he persisted in asking, that this terminal situation was caused by greenhouse gas emissions?

This, I knew, was the wrong question to pose. As Martin Chamberlain had to admit, it wasn't possible to prove definitively that these bears perished because of climate change. But Al Gore did not claim that it was. Instead he had provided a justifiable illustration of the very real threat faced by *Ursus maritimus* as the Arctic sea ice continued to retreat, evidence for which would continue to grow with further research.[32] They hunted from the ice and without it they would starve if they did not scavenge round human settlements or else risk drowning looking for ice floes. Did the movie accurately depict the impacts of climate change if we continued emitting greenhouse gas? That was the question the judge should have asked.

At 4 p.m. proceedings were adjourned. The hearing would resume the following day but I wasn't required in person. My day in court had been a sobering education in the subjective realities of the British establishment. I had seen the first ever application of an iconic piece of politically charged Thatcher legislation, originally drafted to counter Conservative fears of lefty teachers and their perceived agendas to promote gay rights and nuclear disarmament. I had watched fine legal minds, schooled in the humanities rather than the scientific method, struggle to interpret the basis of scientific findings. And I had witnessed a High Court judge subtly tilt the ground on which this particular battle would be fought. This was an individual who had chosen not to view the movie even when he knew it was going to come up in a case before him, or indeed, as he told us during that day in court, taken the trouble to take a look at Al Gore's book of the film, even though he had been given it as a Christmas present the December before.

On the case before him, the judge had decided to focus on the blame game. Only if the scientific community had demonstrated that the examples of disasters depicted in *An Inconvenient Truth* were clearly attributable to human-induced climate change did he appear to accept Gore's illustrations of the effects of climate change. Gore had taken the loss of life in New Orleans during Hurricane Katrina, the drowning of four polar bears in the Arctic Ocean and the desecration of the world's coral reefs as portents of future climate breakdown if we didn't change our ways. The judge chose to see them differently. Because scientists hadn't yet proved, beyond a shadow of a doubt, that the deaths of some people, animals and ecosystems could be directly blamed on emissions of greenhouse gases so far, it appeared that any attempt to portray the likely future consequences of further emissions was open to question and should not be put forward as educational material.

The Judgment was delivered on 10 October and became the lead story on *The World Tonight* on BBC Radio 4.[33] The film could still be shown in schools. But that was not the headline. The 'errors' in the film were. Mr Justice Burton found, 'I have no doubt that Dr Stott, the defendant's expert, is right when he says that "Al Gore's presentation of the causes and likely effects of climate change in the film was broadly accurate."'[34] But, he also found that the law would have been broken had the case not been brought to court and the guidance notes not changed as a result. Their original version had failed to counter the 'more one-sided views of Mr Gore'.[35] They had failed to correct nine of the 'errors' identified by Robert Carter.

Not all of Carter's twenty 'errors' were accepted as such by the judge. But almost half were: the 'Armageddon scenario' of metres of sea level rise, the evacuation of Pacific atolls, the shutting down of the 'ocean conveyor' in the North Atlantic, the link between carbon dioxide and temperature, the melting snows of Kilimanjaro, the drying out of Lake Chad, Hurricane Katrina, the death of polar bears and the destruction of coral reefs.[36] None of these claims, according to his Judgment, could be attributed to human-induced climate change. All were 'significant planks in Mr Gore's political argumentation' that were not corrected in the guidance notes in such a way as to avoid the promotion of partisan views as required in Sections 406 and 407 of the Act.[37]

It was a deeply depressing conclusion. The guidance notes distributed to schools had provided all the context needed for pupils to understand that polar bear drownings, Hurricane Katrina, coral reef bleaching and all the rest were valid illustrations of the effects of climate changes even though scientists had not yet shown to what extent recent events like these could be blamed on emissions of greenhouse gases so far. Since then the effects of climate change have been demonstrated even more clearly on hurricane intensities, on polar bear mortality and on coral reef bleaching.[38] The risk of many metres of sea level rise from the melting of polar ice has risen, and even though it would take centuries, new evidence shows it could be irreversible. There is now evidence that the 'ocean conveyor' in the North Atlantic is slowing down, and although a collapse is very unlikely, a substantial weakening is a real possibility.[39]

But at the time, as I was legally required to, I had answered the questions put to me as best I could. I had tried to interpret faithfully the latest IPCC report and the other supporting scientific evidence I could find. But, despite doing my best on difficult ground, the decision rested on the findings of nine supposed errors that, without the actions of the court, the judge concluded would have led to the illegal indoctrination of schoolchildren.

Thanks to this Judgment, people were now hearing that the science was too uncertain, that *An Inconvenient Truth* was full of errors and that global warming could be ignored. Two thirds of Dimmock's costs were ordered to be paid by the taxpayer, despite his backers having deep pockets.[40] They didn't appear with him on the front steps of the Royal Courts of Justice when the verdict was announced. But they didn't need to. The cheering crowds were elsewhere, in the back rooms of business and politics. They had won another tactical victory in their long-haul campaign for obfuscation of scientific findings and delays in climate action.

Two days later, on 12 October, came very different news. I returned to my desk from attending a scientific seminar to discover that Al Gore and the IPCC had been jointly awarded the 2007 Nobel Peace Prize.[41] This was unexpected but tremendously exciting. I felt honoured to share in such a prestigious award as a member of a team made up of thousands of scientists, administrators and civil servants who had all laboured together to produce a series of increasingly influential assessments of our state of knowledge on climate for the good of people everywhere. Thanks to the carefully prepared and internationally endorsed reports from the IPCC, the world knew enough to act on climate. And thanks to the popular and accessible presentation by Al Gore, probably still the best-known documentary on climate change, an increasing number of people appreciated the urgency of confronting the climate threat. Together, the IPCC and Al Gore seemed worthy winners of perhaps humanity's greatest honour. Being able to share in that success gave me great pleasure.

But it had been a rum week. Halfway through it, I had felt thoroughly deflated at what seemed like a significant setback to the cause of building a wider appreciation of the perilous state of our climate and the urgent need to start reducing our emissions. At its end, I could celebrate a major recognition that the climate threat *was* starting to be taken seriously. Looking back, however, those bittersweet sensations of a tumultuous few days might have served as a harbinger of troubled times to come.

Winning the Nobel Peace Prize had raised the profile of dealing with climate change. But it had also made the scientists responsible for raising the alarm a much greater target for abuse. Climate science was soon to be attacked in the most vicious way yet.

8

Stolen emails

In late 2009, like many people around the world, I thought that governments were about to take the next vital step towards reducing emissions of greenhouse gases. This was supposed to happen at the Fifteenth Conference of the Parties to the UN Framework Convention on Climate Change (COP15) scheduled to take place in Copenhagen from 7 to 18 December 2009. Two years earlier, the Intergovernmental Panel on Climate Change (IPCC) had published its strongest assessment yet. The scientific imperative for decisive action was clear.

And the auguries looked good. Barack Obama was newly installed as US president with a policy of acting on climate change, and many other developed countries were prepared to offer substantial reductions in their emissions. Developing countries, many of whom were most vulnerable to the effects of climate change, were desperate to see progress. In the lead-up to the conference, the widespread expectation was that governments were about to succeed in agreeing reductions that, unlike those agreed in Japan, would keep the rise in global temperature to a relatively safe level.[1]

Less than three weeks before the conference began, the mood music changed. On Tuesday 17 November 2009, NASA scientist Gavin

Schmidt, who ran a website on climate science called RealClimate which sought to rebut the spurious arguments of climate deniers, noticed that an unknown hacker had succeeded in uploading data to his server.[2] As well as the data, a draft post had appeared, stating: 'We feel that climate science is, in the current situation, too important to be kept under wraps. We hereby release a random selection of correspondence, code and documents. Hopefully it will give some insight into the science and the people behind it.'[3] They had included a link to a server in Russia from which the data could also be obtained.

The stolen data was a cache of emails belonging to Phil Jones from the Climatic Research Unit (CRU) at the University of East Anglia (UEA), who were promptly alerted by Schmidt about their security breach. But by now, the climate denier community had also been notified. About the same time as the RealClimate hack, a comment appeared on the website of prominent hockey stick critic and former mining consultant Stephen McIntyre stating that 'A miracle has happened'.[4] Another climate sceptic blog called *Watts Up With That?*, run by a California weatherman called Anthony Watts, was also contacted by the hacker, as were several other climate change denial websites.

I had collaborated with Phil Jones for many years: we published papers together as part of the International Detection and Attribution Group, worked together on the IPCC, and cooperated in compiling data sets of global temperatures. I knew his work well. And like Ben Santer before him, I could see why climate deniers had singled him out for attack. Whereas Ben was the pioneer who had identified the fingerprints of human influence in weather balloon data, Phil was the pioneer who collated weather station data from across the world to produce the most advanced climatic record of global warming. Like Ben, other investigators had later confirmed his findings. And, like Ben, he was about to find out there was a heavy price to pay for his discoveries.

Phil's emails became widely available two days later, on the Thursday of that week. McIntyre was forwarded an internal email from UEA warning university staff that 'climate change sceptics' had obtained a 'large volume of files and emails'. It confirmed to him that the stolen cache was genuine.[5] By now, several denier websites, including the *Watts Up With That?* blog, had downloaded the files and were trawling through them, encouraged by the knowledge that they really did come from UEA. Nevertheless, it seemed that the hacker was getting anxious that the deniers weren't making more of it. The flatmate of one of the moderators of *Watts Up With That?*, a regular poster to various climate sceptic websites called Steven Mosher, received a posting apparently from the hacker claiming that nothing was happening. 'A lot is happening behind the scenes,' Mosher replied. 'Much is being coordinated among major players and the media.'[6] It didn't take them long. That evening I, like many other climate scientists, first became aware that the denial sphere were crowing that they had a scoop: that they had a cache of emails that, they claimed, revealed a dirty truth about the conduct of climate science.

When I came into work the following morning, lurid claims about the emails were swirling across the internet. The emails showed that climate science was rotten to the core, so the story went, that they proved that climate scientists were knee-deep in fraud and deception. As soon as possible, just as Phil was doing over in Norwich, I needed to establish what had been released into the public domain. Journalists were ringing our press office asking to know whether I was standing by my colleague. How could I possibly comment if I hadn't seen the evidence for myself?

A couple of colleagues and I set to work ploughing through the thousands of emails. It took us most of the day. Not surprisingly, this stolen cache of incoming and outgoing emails showed no evidence of inappropriate wrongdoing. Instead, it documented the largely innocuous day-to-day business of the scientist to whom they belonged as he went about his research. At times, Phil and some of his respondents used colourful language to criticize the antics of the climate deniers. Expressing frustrations with their disruptive tactics and misleading statements was not unusual in evening gatherings during conferences, although it was unusual to see such frustrations broadcast like this for all the world to read. But writing them down in emails that were never supposed to be seen by anyone other than the intended recipients hardly seemed like a hanging offence.

Nor did the use of two phrases that had been employed in one particular email. It was obvious to me and anyone who knew about the context that they didn't show anybody doing anything unethical. But they did, I realized when I thought about it travelling home that evening, provide cause for concern about how they could be wilfully misinterpreted. Unfortunately, these two fateful phrases – 'hide the decline' and 'the *Nature* trick' – came loaded with the nasty potential for devious misuse by lobbyists set hell or high water on promoting the cause of rabid climate change denial. They could be twisted to imply that Phil had hidden an actual decline in global temperatures contrary to claims of global warming, and that Phil had used a nefarious trick to present bogus information. And they would, it turned out, be the two phrases that would reverberate around the world for months to come and define the controversy that became known as Climategate.

Over the following days, the confected scandal grew and grew, breaking out of the blogosphere to take centre stage in the mainstream media. 'The Big Climate Change Fraud' trumpeted the front page of the *Daily Express*.[7] 'The worst scientific scandal of our generation', opined Christopher Booker in the *Sunday Telegraph*.[8] The controversy featured in television debates around the globe. On the BBC's flagship political panel programme *Question Time*, *Daily Mail* columnist Melanie Phillips quoted from the emails to support her contention that global warming was a scam. 'There is no evidence for global warming,' she said. 'The seas are not rising, the ice is not decreasing. The temperature is going down not up.'[9]

The Copenhagen summit was looming, and with it the possibility of a new accord that would set the world on a path to reducing its greenhouse gas emissions and avoid the worst effects of global warming.[10] The unfortunate owner of the stolen email cache had done more than anybody else to demonstrate the existence of this warming. Phil had taken weather station measurements, turned them into the most advanced climatic record of land temperature changes and had worked with us at the Met Office to combine them with ocean temperatures to produce the iconic data set known as HadCRUT (which stood for Hadley Centre Climatic Research Unit Temperatures).[11] More than any other set of numbers, HadCRUT had become known for its devastating demonstration of the awful reality of global warming.[12] Phil had also combined instrumental temperature data with other records like the width of tree rings to calculate how temperatures had changed long before the invention of the thermometer. He was a true scientific pioneer.

Stretching back many centuries, the data sets Phil had helped compile showed that the earth was now moving into dangerous territory. If greenhouse gas emissions carried on their current path, we were heading towards a perilous destination that could only be avoided through the sorts of measures set to be adopted in Denmark the following month. These were measures that would protect the environment but could damage fossil fuel interests. It looked like mysterious hands, dancing to the tune of climate change denial, had picked on the scientist most responsible for demonstrating the reality of global warming.

In the run-up to the Copenhagen meeting at which governments, including America's, could sign up to reductions in greenhouse gas emissions, there were plenty of lobby groups who would have been desperate to see some sort of scandal forestall any progress on regulating the growth of fossil fuels. As well as the George C. Marshall Institute, there were many other libertarian think tanks in the US actively trying to prevent progress in environmental regulation. They included the Competitive Enterprise Institute, which in 2005 funded television commercials extolling the benefits of carbon dioxide with the tag line 'They call it pollution. We call it life', and the Heartland Institute which in 2007 funded Fred Singer to write a report by the Nongovernmental International Panel on Climate Change titled 'Nature, not Human Activity, Rules the Climate'.[13]

In recent months, in what looked like a coordinated campaign of intimidation, Phil had been bombarded with a large number of requests to release data he was not able to. In the light of all this, and knowing how important Phil's work had been in demonstrating the reality of global warming, it didn't seem that surprising he had been targeted. I didn't know whether the email theft was a sophisticated hacking attack or an opportunistic grab on a server with lax security. But there could be no doubt that promoters of climate change denial were somehow involved, at the very least through their eager dissemination of the stolen material once it became available.

Collaborating together for so long made Phil's predicament seem all the more agonizing. Together we had worked tirelessly to convince policymakers and governments around the world that global warming was real, that it was a threat to all life on earth and that it was the result of human actions. Assessment, research and evidence: we had travelled that road together, often in the face of stiff resistance from hostile critics in governments, lobby groups and newspaper offices. But all that seemed as nothing to what Phil was facing now.

Thanks to those six words, 'hide the decline' and 'the *Nature* trick'. Even though those words were as innocent of malice as they were loaded with the potential for misinterpretation. And to explain why, you need to understand something about trees.

*

In recent decades, trees have been behaving strangely. For much of their sometimes exceedingly long lifetimes, trees have been very useful for measuring climate. This is because the width of their annual growth rings has served as a reliable proxy of atmospheric temperatures. Warmer temperatures cause greater growth each growing season and therefore wider rings. Conversely, during the years when temperatures are cooler, trees develop narrower rings. This relationship means that tree rings can be used to work out how temperatures changed in ancient times, before direct measurements from thermometers became available. But in the last few decades, trees have changed their behaviour, and this complicates their use for studying climate.

For much of the time of overlap, when we have both tree rings *and* thermometers, variations in temperatures agree well with each other when measured the two different ways. In about 1960, the two methods started to diverge.[14] Whereas temperatures measured by thermometers continued to increase, those derived from tree rings started to decline. It is not clear exactly why, although there are a number of possible hypotheses for what is causing this divergence. It could be that trees are growing less quickly due to warming-induced drought or that reductions in sunlight from air pollution are stunting their growth. With the scientific uncertainties involved in characterizing the exact behaviour of individual trees, it is hard to know exactly why. But what is crystal clear is that it is much better to use direct measurements of temperatures made by thermometers when you have them.

Any proxies of climate, like the widths of tree rings for temperatures, are prone to vary for many more reasons than the one for which you are using it as a proxy. A tree gets sick. Other trees grow up alongside it. Somebody builds a road next door. All these and more will affect its growth. A tree's function is to grow and propagate, not measure temperature. That is what a thermometer was designed to do, after all.

It makes perfect sense therefore to use thermometers to monitor climate when you have them and only use trees and other proxy measures, like ice cores, when you don't. This is the perfectly sensible strategy Phil was describing when he said he was going to 'hide the decline' and use 'the *Nature* trick'. He used these phrases when he was asked to produce an illustration for the cover of a report being produced by the World Meteorological Organization (WMO). This was an annual update on the state of the climate produced every year by the United Nations body responsible for monitoring weather and climate. For 1999's version, WMO wanted Phil to produce a clear way of visualizing how temperatures had changed over past decades and centuries.[15]

Phil presented them with a graph of temperature variations measured in different ways, from tree ring data which stretched back to 1,000 years ago and from the HadCRUT data set which was based on instrumental measurements and which began in 1860. He could have shown the tree ring data even when they appeared to indicate cooling, but that would have made no sense given the known problems in recent years with using tree rings as proxies for climate. Instead, therefore, for well-justified reasons, Phil chose to 'hide the decline', as he put it in that fateful email sent on 16 November 1999. Rather than show the erroneous cooling from trees, he used what he described in the same email as 'the Nature trick'.

'The Nature trick' was a laconic reference to a technique – or trick of the trade – that American scientist Michael Mann had employed when combining instrumental and proxy data in his groundbreaking Nature paper, the one that introduced what became known as the hockey stick and which Putin's right-hand man, Andrei Illarionov, tried to stop me explaining at the bizarre show trial I attended at the Russian Academy of Sciences in July 2004.[16] The 'trick' was simply to cut short the curve of tree ring temperatures when they stopped being reliable. The finished product that appeared on the cover of WMO's State of the Climate report for 1999 was a simplified but accurate presentation of the current state of knowledge on past temperature variations. That was the extent of what the deniers believed was some kind of insidious conspiracy to force the climate change agenda on people.[17]

Ten years on from that long-forgotten brochure cover, Phil was now getting no thanks for it. In the run-up to the crunch meeting in Copenhagen, much of the media attention was being focused on the email controversy rather than on the need to address the climate crisis. Libertarian writer, James Delingpole, wrote in the Daily Telegraph that Climategate was 'a blow to the anthropogenic global warming lobby's credibility from which it is never likely to recover'.[18] And on 23 November, less than a week after the initial dissemination of the emails, a brand new lobby group was founded in the UK to promote climate change denial, headed by ex-Chancellor of the Exchequer, Lord Nigel Lawson. The first act of the Global Warming Policy Foundation was to call for a high-level independent inquiry into the affair.[19] Writing in *The Times*, Lawson claimed that the reputation of British science had been seriously tarnished. 'The integrity of the scientific evidence', he wrote, 'on which not merely the British government, but other countries, too, through the Intergovernmental Panel on Climate Change, claim to base far-reaching and hugely expensive policy decisions, has been called into question.'[20]

Phil, struggling to cope under the pressure, fell unwell and took leave of absence on medical grounds.[21] Phil's employer, the University of East Anglia, made the situation worse by failing to respond to the damaging allegations being hurled in their staff member's direction.[22] To counter the disinformation, the Science Media Centre, an organization dedicated to providing accurate and evidence-based information to the media on controversial science topics, collated responses from a group of climate scientists from across the UK who all reiterated the central point that the emails didn't change the physical properties of carbon dioxide nor the fact that human activity was warming the planet.[23] At the Met Office, we responded robustly to media queries about allegations that global warming had been falsified, and organized an open letter from the scientific community affirming our utmost confidence in Phil's integrity.[24] But Phil's stolen email cache contained many thousands of words and climate deniers were busy trawling through every single one. There was still more devilry to be done, they hoped, in distorting the meaning of more of those individual emails.

I remember well receiving one of the emails they picked out, which had been sent to Phil and a group of colleagues by senior climate scientist Kevin Trenberth from the National Center for Atmospheric Research in the US. In it he rather colourfully described his frustration with the remaining difficulties in measuring the heating of the deep oceans. These were the same difficulties that we had debated at great length in writing the last IPCC report, related to the challenges in accurately measuring seawater temperatures far below the surface. I sympathized with Kevin's frustrations. But Trenberth's words that it was a 'travesty' that we didn't know more at the time were seized on by Melanie Phillips during her appearance on BBC's *Question Time* as evidence of a cover-up by scientists. She claimed we were trying to make inconvenient facts fit our bogus scientific theory of global warming. In reality, Trenberth was simply saying that he thought it was a travesty that more effort and funding hadn't been devoted to reducing this particular aspect of scientific uncertainty. The recipients of his email knew what he meant. Melanie Phillips didn't. But none of us were invited on *Question Time* to debunk her baseless accusations.[25]

It was hard to gauge how all this noise and fury would affect the progress of the climate negotiations in Copenhagen. The week before they began, the chief Saudi negotiator told the BBC they would have a huge impact and could deter countries from offering emissions cuts. 'It appears from the details of the scandal that there is no relationship whatsoever between human activities and climate change,' he said. 'Climate is changing for thousands of years, but for natural and not human-induced reasons. So, whatever the international community does to reduce greenhouse gas emissions will have no effect on the climate's natural variability.'[26] Other countries shared this view, he claimed, bolstering his argument that governments would not agree anything that would affect economic growth until new evidence emerged to settle the scientific picture.

Most countries did, contrary to the Saudi claims, recognize that the scientific evidence for human influence on climate was stronger than ever and demanded action. Even so, there was a worry that suspicions of a cover-up, no matter how far-fetched, had the potential to do damage to the negotiations. Gavin Schmidt's RealClimate website did an excellent job of rapidly rebutting the many specific allegations made about the emails, including explaining why Trenberth's email didn't show scientists hiding an inconvenient truth of global cooling.[27] But there was another issue with the potential to cause even more damage to public trust than everything that had gone before. It was to do with the many requests for data that Phil had received. And it was one that only the Met Office could do something about.

Some of Phil's emails expressed his frustration at the repeated requests he had received to release data that went into compiling his record of temperatures over land. They were requests he had no choice other than to refuse, but his apparent intransigence was now being used to bolster claims of a cover-up. To his growing number of critics, his refusal showed he really did have something to hide, that behind the locked doors of his institute in Norwich he had been falsifying the temperature record to show a level of global warming that wasn't there.

In five days during July 2009, Phil received sixty requests under the Freedom of Information (FOI) legislation asking for the weather station records underpinning HadCRUT. Bearing all the hallmarks of an orchestrated campaign, they were constructed from similar text. One applicant had forgotten to customize theirs which still contained the pro forma instruction provided by hands unknown to 'insert 5 or so countries that are different from ones already requested'.[28] While it was fast work for the requestors, it was labour-intensive for Phil.

Even so, Phil could still have buckled down and got on with releasing the data, recognizing that being unable to perform his normal duties was the price to be paid for rebuffing these serial provocations, were it not for the fact that he was not actually at liberty to do so. Instead, he had to reply to every FOI request setting out why he could not release the data and then respond to the inevitable appeals by further justifying his refusal. The more he dug in, the more he provoked requests for information. Through no fault of his own, he had been put in an invidious position.

Phil had only been able to compile a truly global record of temperature changes by taking measurements from countries who did not allow them to be publicly released. Many countries jealously guarded their weather data which had taken a lot of money to collect. Meteorological stations had been set up and maintained by trained technicians with specialist staff paid to record and archive the measurements. That investment could be repaid, these countries thought, by companies whose operations were sensitive to weather buying this information in the hope that this would help them gain an advantage over their competitors. These were countries who saw weather data as a commercial asset rather than a public good.

Over the years, Phil had got around these restrictions by writing to personal contacts in the countries concerned. These collaborators had allowed him to include their data in his compilations of global climate records provided he didn't send their numbers on to third parties. When he was bombarded with FOI requests, Phil had tried to get them to change their minds, but most were still sticking to their original confidentiality agreements. This information, no matter what its value for understanding the global threat of climate change, was not for redistribution from countries who feared open access could undermine their own domestic markets in weather data. Now Phil's forced refusal to release data was being treated as further evidence of conspiracy and cover-up. We felt we had to do something. But if so, what?

A week after we found out about the stolen email cache, the Met Office chief scientist, Julia Slingo, convened a crisis meeting. We had reached a possible turning point, she realized, in public acceptance of our science. Let this fester and we risked irreparable damage to public trust in the world-leading work being carried out at the research institute she led and our partner institutes, not just in Britain but internationally. I agreed. It wasn't just public acceptance of my own work at risk if this carried on much longer, it was also the work of many colleagues across the world. It was crazy that previous agreements not to release weather observations were undermining international efforts to tackle climate change. If the meteorological organizations of other countries weren't prepared to release their data, we had to get them to change their minds.

Julia had the power to make this happen and hearing her set out her plan at the end of the meeting felt like a huge fillip. She would ask our chief executive to write a letter to every country whose meteorological agency had restrictions on data asking his counterpart leader to give permission for his meteorological service to publish the records from their country going into the HadCRUT data set. We could not expect every country to respond positively to our request. But if enough did, we could compile a new global temperature record consisting only of those countries for which we could release the weather observations, and place this online alongside the computer codes that had created it for everyone to see. For the first time, people could follow for themselves the recipe for creating a global climate record and compile their own record of global warming. Democratizing climate science like this, I hoped, would encourage more people to feel they could have a hand in uncovering and addressing the problem of climate change.

Over the next few days, I felt the fog of despair engendered by the crisis beginning to lift. At last we were doing something positive. From abroad, we were starting to get positive responses to our requests to publish the weather data. Members of my team set to work collating the observations and preparing to recalculate global temperatures using just the publicly available information.

The start of the Copenhagen climate conference, scheduled for 7 December, was approaching fast. We wanted to release our new data set in time to inform the negotiations, ideally before they began, but we also had to give as much chance as possible for heads of overseas meteorological agencies to reply to our chief executive's request. And we had to finish recoding the software. Until we did, and we had carefully crunched the new numbers, we wouldn't know exactly how the new results would turn out, and how closely they'd agree with our previously published results. It was a tense and frantic time.

In Copenhagen, the congress centre where the negotiations

would soon take place was being prepared to welcome thousands of government representatives, lobbyists and journalists to what was set to be the most important climate conference in history. In only a few days the circus would begin. All the while, we continued the race to finish our own preparations. As well as finalizing the code and preparing the data we were going to release, the team wrote a set of twenty-six frequently asked questions and answers to describe the innermost workings of the data set and explain why it hadn't been possible to release the underlying weather station data before. It was another example of the extra work such transparency demands. All this activity took us away from our core research to make the global climate data sets even better, particularly at depicting regional changes. But in the current crisis, it was work we knew was well worth doing.

On 1 December, Phil Jones was asked to stand aside from his role as director of the Climatic Research Unit, and on 3 December, the University of East Anglia announced the setting up of an inquiry into the conduct of the Climatic Research Unit chaired by a senior civil servant, Sir Muir Russell.[29] The inquiry would investigate whether the hacked emails demonstrated manipulation or suppression of data, whether their work complied with scientific best practice, and whether staff had complied with Freedom of Information requests. By Saturday 5 December, we had received sufficient positive replies from meteorological agencies overseas and were far enough advanced with our coding to make our plans public. We announced that, early the next week, we would publish station temperature records for over 1,000 of the stations that make up the global land surface temperature record.[30] We would also show how similar the new results were to the data set that included all the weather station data whether publicly available or not.

On Monday 7 December, in freezing temperatures, the negotiations began. Despite the stresses of recent days, I felt hopeful they'd conclude two weeks from now with a triumphant communiqué of success, as had happened in Kyoto in 1997. But that afternoon, as we were putting the finishing touches to our data release, we heard that Met Office staff who had gone to Denmark to present Hadley Centre research had been standing in line for hours in numbing cold waiting to gain access to the congress centre.[31] The meeting was already in chaos as the organizers struggled to cope with the exceptionally large number of attendees who had arrived for such an iconic event at a venue, it turned out, that was too small.[32] We told our colleagues to go back to their hotels to warm up. They could try again the next day.

More positively, the next day our web page went live with the new information. Changes in global average temperature calculated with just the publicly available weather observations differed little from changes in global average temperatures calculated with all the observations. This was good news, although if I had been thinking more clearly, it should not have come as a great surprise. Phil had already shown ten years before that it takes fewer than 200 well-spread, well-maintained weather stations to get a pretty accurate estimate of global land temperatures and this had recently been confirmed with the latest data.[33] Due to the fact that atmospheric temperatures vary in a consistent way over large geographical areas, it doesn't take observations from that many places to figure out the global average. Removing weather observations that were not publicly available could have only had a small effect on the overall global temperature trend.

The important thing is to select weather stations that can be trusted for climate monitoring. This task was a large part of what Phil Jones had devoted his career to. Over time, weather stations can change their location from city centre to out of town airport, or they can stay where they are while their surroundings change, as for example suburbs spring up around them. Such non-meteorological changes can alter the readings in ways that have nothing to do with global climate change. Phil worked out how to tell which were the reliable stations to trust for climate monitoring by comparing changes in one location with those at a nearby site: sudden differences springing up between them could mean that one or the other had a problem and should be corrected or else eliminated from the record. By building up a database of reliable, well-maintained weather stations across the globe, Phil enabled accurate measures of global temperatures over land to be constructed, not just as a global average but as maps showing which places were warming up more than others.[34]

The full power of Phil's work was made evident when we combined his record of temperature over land with how temperatures had varied over the oceans. This was the specialism of my Climate Monitoring and Attribution team at the Met Office who had compiled a record of changing temperatures over the surface of the seas.[35] Combined with Phil's land temperature data, they produced an iconic record of global temperatures over both land and sea, the results available on the Met Office website for all to see as HadCRUT.[36] It showed us what emitting greenhouse gases since the Industrial Revolution had meant: a warming of about 1 degree Celsius over the last hundred years.

Now, at the start of the Copenhagen conference, we had a new version of global temperatures made up of freely available data. As a result, while the negotiators in Denmark had a host of tricky issues to try and resolve, at least they didn't need to worry about the provenance of global temperature records as the Saudis had claimed. And with the publication of code to make their own global average temperature time series, people would be able to verify for themselves that the world was indeed warming, just as we experts had said all along.

We had blocked off one line of attack from climate deniers. But if we'd hoped to boost politicians' morale ahead of the crucial climate talks, there were other factors at work darkening the delegates' mood in Denmark. The weather was terrible, with unusually cold temperatures and heavy snow. And the organization was shambolic. It wasn't just accredited delegates who were poorly treated. An idealistic son of a friend of mine cycled all the way from Britain to protest about the lack of action on climate change. Part of a peaceful demonstration with thousands of others, he and the group of protesters around him were subject to an unprovoked assault by police, kettled for hours in the freezing cold, then bussed to a hall where they were further detained without charge or access to food, water or toilets.[37]

After two weeks of deadlock, the political negotiations ended in acrimony late on the final night. There had been a complete breakdown of trust between developed and developing countries after proposals being worked on by the former group were leaked. And when Obama arrived during the final hours of the conference, he convened a meeting of some leading nations – including Brazil, South Africa, India and China – but excluded others, to create the Copenhagen Accord. This proposal recognized the case for keeping temperatures below 2 degrees Celsius above pre-industrial levels but did not include the emissions reductions that would be needed to do so. It proved unacceptable to many of the countries not involved in drawing it up.[38]

I noticed a striking photo in the Sunday papers.[39] It showed the most powerful leaders of the Western world sitting round a small table in the lobby haggling over a deal that had already gone sour. An interpreter next to the French president is translating Merkel as she leans forward towards Obama, her hand outstretched with downturned palm in a gesture of refusal. Obama is looking at his watch. As though he has been designated secretary of the group, Prime Minister Gordon Brown, apparently being ignored by everybody else, is taking notes.

Despite leaders like Obama and Brown trying to talk up the small steps that had been made towards agreeing legally binding emissions reductions at some later date, no climate scientists saw 2009 as having ended in success. The chair of the IPCC, Rajendra Pachauri, trying to be diplomatic, stated that the Copenhagen agreement was 'good but not adequate'.[40] Environmental groups went much further in their condemnation. According to John Sauven, the executive director of Greenpeace UK, 'The city of Copenhagen is a crime scene tonight... It is now evident that beating global warming will require a radically different model of politics than the one on display here in Copenhagen.'[41]

The failure of politicians to act on our scientific advice which itself had come under vicious attack felt like the lowest point for climate science since those first groundbreaking studies of 1996 by Ben Santer and Gabi Hegerl that identified the human fingerprint of climate change in atmospheric temperature changes. Since then, our community had found more and more evidence for the effects of greenhouse gas emissions on the oceans and atmosphere, had linked emissions to extreme weather events like the European heatwave of 2003 and had shown that continued emissions would lead to many further changes in climate including retreat of snow and ice, rising sea level and more intense storms.

We had told the world what was happening and for a time it had felt like the world was taking note. The Kyoto Protocol of 1997 had made a start in signing up countries to reductions in greenhouse gas emissions. The release of the findings of the IPCC report in Paris in February 2007 on the physical science basis of climate change had seen an extraordinary gathering of journalists. It was probably the largest media turnout ever for an announcement on climate science. And there was another report later that year that was also influential, even if it didn't receive anything like the same attention in the press.

This was the IPCC's Synthesis Report, a summary document that combined findings from across the three working groups of the IPCC and included assessments of the impacts of climate change and the reductions in greenhouse gas emissions needed to avoid the worst effects.[42] This report, released at a meeting in Valencia, Spain in September 2007, laid out the risks if climate change continued unchecked. In it, five 'Reasons for Concern' were identified. These were: the risks to unique and vulnerable systems, including Arctic communities and coral reefs; the risks of extreme weather events; the uneven distribution of the impacts of climate change, with those in the weakest economic position suffering the most; the aggregate costs of climate change; and the risks of large-scale singularities, such as the collapse of the Greenland and West Antarctic Ice Sheets. The report found that these risks increase substantially once the rise in global temperature reaches 1.5 to 2.5 degrees relative to pre-industrial levels. At above 4 degrees of warming, the risks become extreme, with the world threatened by widespread extinctions, food shortages and inundation of coastal regions. To keep warming to below 2 degrees, the report concluded that substantial reductions in emissions were needed, of at least 50 per cent by 2050 relative to current levels.

The award of the Nobel Peace Prize later that year to Al Gore and the IPCC seemed to underline the significance of the warnings we'd been giving. Governments stood on the threshold of upping their ambitions on climate and ready to make definitive steps away from the dangerous future that awaited everyone on earth if emissions continued.

In the UK, the 2008 Climate Change Act had mandated targets for emissions cuts according to advice from the independent Climate Change Committee. In their first report, the committee advised the government that the global climate objective in line with the IPCC's latest assessment should be to keep warming to no more than 2 degrees above pre-industrial levels.[43] This would mean reducing greenhouse gas emissions globally by 50 per cent below 1990 levels by 2050. Recognizing that developed countries would need to do more if their emissions in 2050 were going to be close to the global per capita average, the committee recommended a UK target of a 34 per cent cut by 2020 and 80 per cent by 2050. In April 2009, in what was the world's first legally binding carbon budget, the government pledged to make 34 per cent emissions cuts by 2020 in line with the committee's recommendations.[44] It meant that the UK was able to argue in Copenhagen that they had done their bit towards a global agreement to reduce greenhouse gas emissions sufficiently to avoid the worst effects of climate change. It felt like the message of science was getting through.

But then it had all gone horribly wrong. The disastrous meeting in Copenhagen showed world leaders looking uncertain and divided. Climate science was being widely vilified. The idea that our scientific research might lead to action one day seemed a distant and forlorn hope.

Like many hundreds of colleagues, I had spent many hours away from home working on the IPCC. The work we did gathering together many thousands of pieces of evidence and compiling our conclusions in a way that was clear and convincing took us away from our own research further advancing the frontiers of knowledge. But we found it necessary and rewarding if governments took note of our findings and made progress in taking action in response. With the prospect of global reductions in emissions now seeming as far away as ever, it was immensely dispiriting that our efforts had so far largely been in vain.

Nevertheless, I was still paid to go to work and there was one more job I wanted to get done before breaking up for the Christmas holidays. A journal paper I had been working on for over a year was nearing completion. It was a substantial piece of work, written with six colleagues from the International Detection and Attribution Group, which reviewed the progress that had been made in understanding the causes of climate change since the last IPCC report was released in 2007. Based on the conclusions of 110 scientific papers, we found that 'further evidence has accumulated attributing a much wider range of climate changes to human activities'.[45]

Thoroughly reviewed by independent experts, it was important to me that this updated assessment was published, whether or not policymakers were prepared to take on board our conclusions. For the record, despite the failure of the climate negotiations and the enervating setback of Climategate, whoever was prepared to listen should know that the evidence for the seriousness of climate change had strengthened considerably, even since the most recent IPCC report of only two years ago. It was my professional duty to finish the work off. But also, after the crushing disappointment of Copenhagen, speaking out about this research of mine felt, during those dark days, like an act of defiance.[46] I would not like history to judge that I had been silent at this critical time.

I worked through the day of Monday 21 December, while thick snowflakes drifted past my office window. Corrected proofs submitted,

I set off for home. With the roads now impassable and with buses out of action, I trudged home for an hour and a half through the snow past stationary cars. People living along the muffled streets were bringing out cups of tea for stranded motorists, small gestures of solidarity in a dysfunctional world.

The winter of 2009–10 was the coldest winter in the UK for thirty years.[47] Sadly, and due in no small part to the wilful obfuscations of the climate deniers, many people remained confused about how such severe cold could co-exist with global warming. In reality, we could trace what our climate models had long predicted, a steady upswing of temperatures globally and locally with upticks and downticks along the way, both cold and hot. There were still record cold temperatures being recorded but they were being outnumbered by record hot temperatures by a factor of ten to one.[48] Occasional winter snow didn't mean the climate heat was off. It was a simple point but little appreciated during that frigid winter.

Christmas was a welcome time to forget about work for a while, take a breather and de-stress. The soothing traditions of Christmas carols, mince pies and a decorated tree made everything seem all right for a while. Time would be a great healer and hopefully Phil Jones would eventually be able to put all this unpleasantness behind him and get back to doing what he did best. By the middle of January, he was back at work and my team and I were collaborating with him on updating HadCRUT with newly acquired weather observations. Phil seemed in better spirits. The accusations of scientific malpractice had quietened down somewhat. It seemed like the worst was over.

The first week of February brought a nasty reality check. 'Key study by East Anglia professor was based on suspect figures', claimed a front page exclusive in Tuesday's *Guardian*.[49] The next day the newspaper followed up with another lead, entitled 'Guardian investigation reveals how scientists kept critics out of journals'. The paper had abandoned their two very knowledgeable in-house correspondents, Leo Hickman and David Adam, in favour of veteran freelance, Fred Pearce, who had a book coming out analysing the hacked email cache. Apparently, according to the progressive paper's editors, this justified a whole week of front page splashes, all of them highly critical of Phil Jones. Pearce claimed that a series of measurements from Chinese weather stations were seriously flawed and that documents relating to them could not be produced.[50] An email from Phil's colleague, Keith Briffa, was dredged up as evidence of distortion of the peer review process.[51] In an accompanying editorial, the paper linked the emails to a recent poll showing an increase in public scepticism in climate change. 'What Copenhagen did for the chances of a meaningful climate deal, East Anglia has unwittingly done for the prospects of prevailing in the battle for hearts and minds', opined the leader.[52]

I'd read the emails. I knew the context. And they didn't show the sort of malpractice the paper was trying to claim. In no way had Phil misused Chinese weather station data nor, despite a peevish email from Keith Briffa, had the peer review process been distorted by scientists from the Climatic Research Unit. Even so, the ecological campaigner and *Guardian* columnist George Monbiot weighed in, calling for Phil's resignation.[53] Previously, it had been those arguing that climate change was a confidence trick that were demanding heads to roll over Climategate. Now, it was those who thought that climate change posed a terrible threat requiring drastic action. For some of them too, Phil was a traitor to the cause.

Disappointing as it was to see climate scientists in the sights of environmental campaigners, the greater threat to our jobs and even our freedom came from the other side of the debate, those who would go to great lengths to ensure continued consumption of fossil fuels. On 23 February, Senator James Inhofe, Republican chair of the Senate Committee on Environment and Public Works, released a list of seventeen climate scientists that he wished to see investigated for possible referral to the US Justice Department for prosecution. According to Inhofe, the CRU emails revealed unethical and potentially illegal behaviour by some of the world's pre-eminent climate scientists. In addition to Phil Jones, the list included many people I knew well, including Susan Solomon, Kevin Trenberth, Ben Santer, Michael Mann and Gavin Schmidt. It also included me.[54]

If Inhofe were eventually successful in his bid to make us outlaws, I could face arrest if I travelled to the US or even extradition from the UK. It was a chilling reminder of the sort of witch-hunt I could face for uncovering facts that were unwelcome to powerful people. Thankfully, there was a Democrat in the White House and Democrats controlled the Senate; Inhofe's motion got nowhere. But the incident served as a sobering reminder of the thin ice on which we climate scientists trod. If one day America had a hard-right, populist, pro-oil administration in power with a supportive majority in Congress, Inhofe's motion could get passed and I and sixteen other of the world's leading climate scientists might face prison.

If such a dystopian future seemed slightly far-fetched in early 2010, the Climategate affair did nevertheless have a chilling effect on the willingness of climate scientists to come out in public and defend their work. Unused to the roughhouse tactics of a re-energized band of climate deniers, most retreated to the safety of their labs and hoped for better times. Talking about your latest research to the media also meant defending your science from attack. Few were prepared to do that, especially when most weren't versed in the media skills needed to handle unfair and biased challenge. As American ecologist Paul Ehrlich put it, 'Everyone is scared shitless, but they don't know what to do.'[55]

In the British Parliament, the Science and Technology Committee of the House of Commons had decided to have an investigation into the Climategate affair. On 1 March 2010, I saw Phil Jones on *News at Ten* sitting next to his vice chancellor, facing the MPs. He looked gaunt and harassed. Forced into an occasion for which he was eminently unsuited, he was sounding unhelpfully defensive. Having his vice chancellor there didn't seem to be helping him much. 'I have obviously written some very awful emails,' he told them.[56] Put somebody under enough pressure, especially if they don't have the right support, and innocent people will eventually admit they must have done something wrong. That's what seemed to have happened to Phil.

Three days later I had an opportunity to step out of the lab and do what few others seemed to be doing: start a fightback on behalf of climate science. My paper assessing progress in attributing climate change since the last IPCC report was published, and the Science Media Centre asked me to brief the media on the new work at the Royal Institution.[57] When I walked into the briefing room that morning, along a dingy upstairs corridor in the venerable but antiquated home of the Royal Institution in central London, I was in for a shock. Rather than just a few reporters as these technical briefings tended to attract, I was faced with what looked like the cream of the UK's environmental and science reporting establishment. From the BBC, I recognized David Shukman from TV news programmes, as well as his BBC colleagues Pallab Ghosh and Tom Feilden. And from the newspapers there were correspondents from the *Daily Mail*, *Financial Times*, *New Scientist*, Press Association, *Independent*, *Mirror* and *Guardian*. A scatter of reporters' microphones had been placed on the table in front of me, like one of those press conferences where the local police chief is giving an update on the latest progress in his investigations into a terrible crime. At the back of the room, a video camera was in position, ready to roll.

Nervously at first, I launched into a brief summary of what my latest analysis showed, published today and based on a review of over one hundred peer-reviewed scientific papers. The reality of global warming was clear, I told the reporters, as could be seen not just from the warming at the surface documented by Phil and our team at the Met Office, but also by the increasing heating of the oceans and the rapidly rising sea levels. Natural variability, I explained, either from the sun, volcanic eruptions or natural cycles, could not explain the warming observed in recent years. It was, however, consistent with the effects of human activity including human-induced greenhouse gas emissions.

Getting into my stride, I went on to lay out where the effects of human-induced warming had been seen. It had been detected on every continent, as well as in the warming of the oceans and in changes in their saltiness. The atmosphere had become more humid as more water evaporated from the ocean surface. Precipitation patterns had changed, with wetter areas seeing more rainfall and drier areas less. And the earth's snow and ice was melting, in the Arctic at a terrifically rapid rate. I summed up. The fingerprints of human influence on climate were everywhere. Based on the latest research, there could be no doubt about the significant and substantial nature of the climate changes wrought by humanity's emissions of greenhouse gases.

With the ten minutes allotted to my introduction used up, it was time for the questions. A forest of hands shot up. And then, with the first salvo in my interrogation, came the next shock of today's surprising turn of events. In my previous experience, such sessions tended to resolve around the implications of the research in question to the everyday lives of their readers and viewers. Not this time. What awaited me now was something totally novel.

They wanted to know about the actual research. David Shukman, the most well-known reporter in the room thanks to his regular appearances reporting on science and environment for the main BBC television news bulletins, set the tone with his opening question. How exactly could we be sure, he wanted to know, that global warming was really down to human causes?

Natural variations of the climate system, I explained, processes like the El Niño–Southern Oscillation, don't produce anything like the pattern of warming that has been observed in recent decades, in particular the greater warming over land than ocean and the fastest warming at the North Pole. Also, while increases in solar output could have caused a small upturn in global temperatures, it played a very minor role compared with that played by greenhouse gases. But why, Shukman pressed on, was it possible to so confidently rule out the sun? Increases in solar irradiance, I elaborated, produce a very different fingerprint of atmospheric temperature changes than increases in greenhouse gases. While the former cause warming throughout the depth of the atmosphere, the latter cause exactly the same distinctive pattern of tropospheric warming and stratospheric cooling as has been observed.

I was giving him essentially the same account that Ben Santer had provided at that extraordinary session at the American Geophysical Union meeting in San Francisco thirteen years before, which had formed my initiation into the contentious subject of climate change. It was a bittersweet irony that I finally had the ear of the media, explaining science they should have known about more than a decade earlier. Then, Ben was responding to politically motivated attacks from lobbyist Patrick J. Michaels. Now, I was responding to a newly minted critical investigation instigated by journalists. Previously, they hadn't been much interested in these technical details. Now it was starting to look like this was all they wanted to talk about.

The next person wanted to know how climate models worked. Someone else wanted to hear more about how we constructed the HadCRUT global temperature record. The meeting was turning into a scientific seminar. I should have felt at home with such questions but it wasn't easy to respond in language non-scientists would understand. But I did my best. People in front of me seemed to keep scribbling at least. And then, from the next questioner, came a change of tack. It was the third shock of the morning.

Was my study the beginning of a fightback against the climate sceptics? It was a question I should have expected. But with all these technical questions I had been lulled into thinking this was a scientific seminar rather than a press conference. I thought for a moment and then answered by giving them the facts.

It was over a year since I had started writing the paper and I had submitted it to the journal long before the release of Phil Jones's emails. I wanted people to look at the evidence I had presented, I said, and make up their own minds about the seriousness of climate change. Feeling drained after the grilling I'd just faced, I left it there. But knowing how reluctant most scientists had been to put their heads above the parapet like this, I hoped that publicizing my paper would be the start of a serious fightback. The nonsense from the climate deniers that had gained such traction in the media and with the public during the last few months needed to be seen for what it was: the fringe views of an ill-informed and tiny minority.

The headlines the next day told the story of how my work had been received. 'After emails and errors,' the *Guardian* titled its article, 'fightback begins to prove global warming is caused by humans'.[58] Underpinning this 'fightback' narrative came accounts of the scientific evidence on which our claims of human influence on climate were based. Thankfully, I had been able to convince the reporters that the research was solid. Later I learnt from newspaper contacts that journalists were under pressure from their editors, most of whom were against climate action. These editors' support of climate deniers for many years and the column inches given to David Bellamy, Nigel Lawson, James Delingpole and others had made it hard for scientists to get their message across in the mainstream media. This was pretty appalling in the light of the mounting evidence that contradicted the deniers' claims and the efforts of their science and environment correspondents to persuade their editors otherwise. Now, in the light of Climategate, many of these editors were accusing their environmental staff of going native with suspect scientists. As a result, the reporters had to justify their continued reporting on the developing climate crisis, not just to their readers but to their superiors. This explained the grilling served up to me at the Royal Institution.

If a fightback really was under way, that was obviously very good news. But I knew that if this fightback was going to endure, it needed a clean bill of health from the various official investigations into Phil's conduct. I awaited the outcome of these reviews with hopeful expectation, but also with some degree of trepidation. Phil had not responded well when under pressure by the Commons committee in the Houses of Parliament. I feared what impatient adjudicators might make of unpolished performances like this.

Thankfully, my worries proved unfounded. The Science and Technology Committee published its report on 31 March. It found that 'the scientific reputation of Professor Jones and CRU remains intact'.[59] The Muir Russell inquiry reported in July. It found that the 'rigour and honesty' of the scientists at CRU were not in doubt.[60] And a third committee of experts, set up by the University of East Anglia in consultation with the Royal Society, to assess the integrity of the research published by CRU, concluded in its report published on 14 April that there was 'no evidence of any deliberate scientific malpractice in any of the work of the Climatic Research Unit'.[61] Three different committees, examining the issues raised by Climategate from all possible angles, had each come to the same conclusion. Phil had been exonerated. This was a huge vindication.

The next year brought further confirmation that the climate science community had been right all along about global warming. The temperature increases seen in the HadCRUT record compiled by Phil Jones and ourselves at the Met Office were also seen in two alternative records from NASA and the US National Oceanic and Atmospheric Administration. But suspicious that Climategate could have revealed a common flaw in all three data sets, a self-appointed task force of theoretical physicists from the University of California at Berkeley decided to investigate.[62] Led by physics professor Richard Muller, the ten-strong team included that year's Nobel Prize-winning astrophysicist Saul Perlmutter and climate sceptic blogger Steven Mosher, and it was funded with a $150,000 grant from the climate change denial-supporting Charles G. Koch Foundation.[63] Going back to first principles, the group took weather measurements from across the globe and made their own calculations of how global temperatures had changed. Their results supported what Phil and his collaborators had been saying for years. 'Our biggest surprise', Muller was reported as saying, 'was that the new results agreed so closely with warming values published previously.'[64]

The result of the analysis must have been a blow to the Foundation, funded by the billionaire Koch Brothers whose Koch Industries comprised a vast enterprise of oil refineries, pipelines and fertilizer facilities. In 2009, the company spent over $12 million on lobbyists, fifth in the league table of oil and gas sector lobbying behind Exxon, Chevron, ConocoPhillips and BP.[65] Charles Koch had also co-founded libertarian think tank the Cato Institute, whose Center for the Study of Science was headed by Patrick J. Michaels and included Richard Lindzen among its 'distinguished fellows'.[66] Following the email release, Lindzen had told an audience at his institute, MIT, that the central issue of Climategate was 'the alleged suppression and manipulation of raw temperature data by climate scientists'.[67] Michaels had told the *New York Times* that evidence for alleged suppression and manipulation of data in the emails was 'not a smoking gun' but 'a mushroom cloud'.[68] How disappointing it must have been to the Kochs that their experts' advice stating that there could be a fatal flaw in global warming records had been falsified.

Phil was reinstated to CRU as director of research in July 2010, allowing him to get back to work analysing the details of weather records for monitoring changes in our climate.[69] He would never send emails to close colleagues the same way again, sarcastic, critical of sceptics, flippant. Powerful people would soon come to terms with the idea that trillion-dollar decisions rested on a handful of researchers funded by soft grant money in places like the Climatic Research Unit. Global warming was unequivocal. Phil Jones and his colleagues had shown it so.

With Climategate behind us, a not guilty verdict having finally been delivered, I felt a renewed sense of purpose about my own mission in climate research. My work was not foolish if it led to a greater understanding of the effects of increasing greenhouse gas emissions and if it helped redouble efforts to find ways to counter the mounting threat of continued warming. What's more, later that year, I received a remarkable opportunity to help translate scientific developments into policy-relevant advice that could inspire action. I was asked to be co-leader of the section on the detection and attribution of climate change in the next IPCC report, the UN body's fifth major assessment, due to be published in 2013. It was an offer I couldn't refuse.

Climate negotiators now hoped to make a fresh start. Willing countries were keen to build a new momentum towards a genuine, and long overdue, global commitment to cutting emissions. To support them, they needed a new scientific assessment, one untainted by the noise and fury Climategate had unleashed. Taking on this responsibility would give me a large role in steering the report and seeing it agreed by governments. It was a prospect I relished.

While climate deniers had been doing their worst and politicians had been prevaricating, extreme weather events around the world had become more and more frequent. Time was of the essence; the number of victims of climate change was mounting. In the course of working on the next IPCC report I would have my first close exposure to the terrible truth of what that meant for some of the people involved.

9

Mounting devastation

The New Zealand newspaper the *Dominion Post* carried a shocking image on the front page of its edition for Wednesday 9 January 2013. It showed a woman and five young children cowering beneath a jetty in a small Tasmanian township called Dunalley. The terrified victims were floating up to their necks in water while red smoke billowed around them. 'Family of 7 escapes tornadoes of fire' was the banner headline that accompanied the picture.[1] According to the article, record-breaking heat and rapidly spreading flames had brought them to this pass. Stranded in this desperate limbo, breathing toxic air while shivering in clammy water, this photo seemed to sum up what climate change had become for many. Global warming was no longer a question of discernible anthropogenic fingerprints or future climate impacts; it was a matter of life and death.

I had picked up the paper while on holiday across the Tasman Sea in the land of the long white cloud, hiking through the gorgeous scenery of the country's South Island. I had been struck by this story, not just because of its graphic depiction of weather-related jeopardy, but because I was soon to travel to close by where the picture was taken. A few days later, I took that long-anticipated trip, my first to

Australia, to attend the fourth and final lead author meeting for the Intergovernmental Panel on Climate Change's (IPCC) Fifth Assessment Report in Hobart, Tasmania's most populous city. As my plane came into land, I could see burnt landscapes around the airport and smoke billowing around the Tasman Peninsula where the small settlement of Dunalley stood.

From my hotel room I could still see clouds of smoke, a pertinent reminder of why what I did mattered. For the first time since scientists had started to gather to prepare IPCC reports in the late 1980s, our lead author meeting was happening in the immediate aftermath of a weather-related disaster. That felt significant.

I knew that around the globe, the toll of heatwaves, floods and droughts continued to mount. But the fact that such devastating fires had struck so recently so nearby made our task here seem all the more significant. Coincidentally, my co-leader of the section of the IPCC report on the causes of climate change, an Australian scientist called Nathan Bindoff, was based here at the University of Tasmania. For him, the local conflagration was not just of professional interest but also of personal concern, given his own home's vulnerability to such fires. The recent events only confirmed a decision we'd made earlier in the writing process: for this IPCC report we needed to break new ground.

Previous reports had concentrated on analysing long-term changes in climate, assessing what had caused the increasing atmospheric temperatures over many decades, the diminishing snow and ice and changing rainfall patterns. It was vital we updated previous assessments with the latest knowledge, so we could determine whether the scientific confidence in the role of human activities on climate had increased. But this time, Nathan and I agreed, we also needed to assess the extent to which extreme weather events in recent years could be linked to human-induced climate change.

Over the days to come, I knew I had to try to keep in mind the big picture, the global-scale changes in climate we needed to attribute, as well as the local – how individual places were being affected by the rising concentrations of atmospheric greenhouse gases. At the end of the week, at the conclusion of our IPCC deliberations, I had agreed to give a public lecture at the university in Hobart. I couldn't help thinking, as I waited for the pressured hectic schedule of our meeting to start on the Monday morning, what I could tell the local people on the Friday evening. I also wondered what I might learn from them.

There were many aspects of what had happened just up the road in that little community of Dunalley that intrigued me and made me want to know more. According to the article in the New Zealand newspaper, on Friday 4 January Bonnie Walker, the mother of the youngsters and daughter of the woman in the picture, had braced herself to lose her parents and five children. With her husband away hiking with friends, Mrs Walker had left her children that morning in the capable care of her parents to drive over to Hobart, 50 kilometres away, so she could attend a funeral. Hearing what had struck back home when she arrived in Hobart, Bonnie Walker feared the worst.

The picture that featured on the *Dominion Post*'s front page and in many other media outlets was taken by Bonnie Walker's father and texted over to his daughter to reassure her that her children were alive, even though they were clearly in dire straits. Her dad, Tim Holmes, and her mother had evacuated themselves and their five grandchildren away from the raging inferno to the only sanctuary they had left. 'There was so much smoke and ember', Mr Holmes told the paper, 'and there was only about probably 200 millimetres to 300 millimetres of air above the water. So, we were all just heads, [with] water up to our chins just trying to breathe because the atmosphere was so incredibly toxic.'[2] After three hours with the children up to their necks in water, the blaze had subsided enough for him to retrieve a dinghy from the shore which allowed them to make their escape around the point in the direction of the Dunalley Hotel. After spending a sleepless night waiting for help, they were found by rescuers the next day.

The whole story left me with so many questions. How could the fire have arrived so quickly that Mrs Walker hadn't realized the impending danger when she left home that morning? How were the record-breaking temperatures that day – 41.7 degrees Celsius – possible in a temperate ocean-moderated climate that was rather like Britain's? And how could it be possible to link this extraordinary chain of events to human-induced climate change, like we'd done with the deadly European heatwave of 2003?

On Monday morning, for us IPCC authors it was down to work. Outside, temperatures had dropped to a breezy 20 degrees, but inside the conference centre where our meeting took place we were up against it if we were going to finish what we needed to achieve. We had five short days to agree on the key findings of our chapter on the detection and attribution of climate change. Over the last two years and three previous lead author meetings, we had assessed the information from 669 peer-reviewed studies. We had examined changing temperatures at the surface and in the atmosphere, the warming oceans, rising sea levels, altered saltiness of seawater, ocean acidification, decreasing oxygen levels in ocean waters, changing precipitation patterns, increasing humidity in the air, glacier melting, melting of the Greenland and Antarctic Ice Sheets, Arctic sea ice loss and melting snow cover. We had abundant evidence to draw on. From tens of kilometres above the earth to several kilometres beneath the surface of the oceans, from the poles to the equator and across all the continents, human activities on climate were causing rapid and devastating changes. Our job in Hobart was to agree on the best way to get that message across as clearly as possible.

To help us do that, we worked on a table that itemized the changes in many different aspects of the changing climate. With thirty-three detailed entries (that would stretch over eight pages of the final published document), it was the most comprehensive collation of evidence for human-induced climate change that the IPCC had yet produced. With that in hand, we next needed to distil the information into a short summary of our most important findings. These were the ones that would be highlighted in the Summary for Policymakers, the crucial section that had to be agreed word by word with governments at an IPCC meeting later in the year in Stockholm. And as we worked on the table and the summary statements, we needed to take account of over 2,000 comments by independent experts on the most recent draft of our chapter, either by adjusting our text or else by explaining why we disagreed. Only by working in such a thorough way could we be confident that our assessment would be supported by the weight of scientific evidence and would be robust to critical scrutiny. We owed it to the governments in Stockholm to whom we would present our conclusions. More significantly, perhaps, we owed it to the many people whose lives and livelihoods, like the Holmes family from Dunalley, were being threatened by a rising tide of climatic disruptions.

It was a hectic, pressured time assessing all this information under the overall direction of Swiss scientist Thomas Stocker who had taken on the leadership role held by Susan Solomon for the previous report. Working alongside Chinese scientist Qin Dahe, Thomas oversaw the main findings being gathered from all the chapters, including those that were documenting the likely effects of climate change if greenhouse gas emissions continued unabated. The evidence there was sobering: temperatures would continue to increase, snow and ice would continue to melt, and sea levels would continue to rise. Even worse, the ability of the oceans and atmosphere to absorb part of the polluting carbon dioxide emissions would be compromised, making climate change even more severe. It all begged the question of what needed to be done to prevent all this.

The answer lay in a striking new scientific finding that all of us gathered in Hobart knew had far-reaching implications. It was the discovery that global warming from carbon dioxide emissions is directly related to the total sum of emissions so far. That means that if the total carbon dioxide emitted by humanity doubles, the rise in global temperature also doubles. Despite the wonderful complexity of our beautiful planet, in this respect at least nature is really quite simple.

This discovery, which was first made by Myles Allen and published in *Nature* in 2009, was later confirmed by further independent studies.[3]

But it was far from obvious. Global temperatures vary from year to year and decade to decade, variations which are not directly in lock-step with mounting emissions of greenhouse gases. Indeed, the paucity of record-breaking warmth for a decade following the record hot year of 1998 became a cause célèbre for climate deniers who claimed that this showed that global warming had stopped, even though global temperatures peaked again in 2010 and then surged ahead in subsequent years. But once you've taken account of the natural variations in temperature, which are only short term and whose peaks and troughs cancel out over time, the strong underlying warming trend has a simple and direct relationship with the total emissions to date. This means that, to avoid global temperatures reaching 1.5 or 2 degrees Celsius above pre-industrial levels, there is little carbon dioxide left to emit because of the large amounts that have already been released.

This would be one of the most important conclusions that we were going to put to governments in Stockholm. Alongside it would sit the IPCC's headline conclusion about the causes of climate change. A key task for our chapter team therefore was to craft a single sentence summarizing the role of human influence on past warming. For the previous report, the one that I had seen agreed by governments in Paris in 2007, the headline conclusion had been that 'most of the observed increase in global average temperatures since the mid-20th century is *very likely* due to the observed increase in anthropogenic greenhouse gas concentrations'. The statement used calibrated likelihood language, meaning that we had employed the term '*very likely*' to indicate that we assessed a greater than 90 per cent probability in the correctness of this statement.[4] Now we needed to provide an updated statement that reflected the scientific advances that had been made over the intervening six years.

In addition to the greater wealth of evidence of changes across the climate system that we had itemized in our thirty-three-entry table, we also had new evidence about the heating of the ocean. Since the previous IPCC report, scientific research had resolved the conundrum that had prevented us agreeing to a stronger attribution statement at the Paris meeting in 2007. The conundrum was in knowing how ocean temperatures were varying, particularly during the 1980s. Previous estimates of ocean temperatures had shown them cooling during this period, a cooling that disagreed with climate models which instead simulated warming.

Now, this cooling had been shown to be erroneous, a result of incorrect assumptions about how fast instruments for measuring the temperature of seawater, expendable bathythermographs (or XBTs for short), fell through the water as they sank into the depths. New updated estimates of ocean temperature changes took proper account of the fall speeds of XBTs.[5] They showed a steady warming over the last fifty years, in line with what climate models had long since predicted. Now that the ocean cooling conundrum had been resolved, one of the vital caveats in the previous IPCC assessment had fallen away. Taken together with the increased evidence of changes right across the climate system, we could now be more confident than we were for the last report about the dominant role of human influence on past warming. This felt like a significant milestone.

Summing up the current state of knowledge, we agreed in our chapter room in Hobart that it was '*extremely likely*' that human influence was the dominant cause of the observed warming since the mid twentieth century. Our increased confidence in the role of human influence was reflected in our use of the phrase '*extremely likely*', which meant there was a greater than 95 per cent probability of it being correct. Our assessment had been carefully justified in the detailed text of our chapter, had been supported by a slew of independent experts who had reviewed our drafts and had been agreed by all the authors of our chapter. This is what we would put to governments in Stockholm.

Reaching such staging posts along the long march to a finished assessment provided welcome moments of relief in a task that threatened at times to become all-consuming. Working all hours when out at meetings was easier to deal with than never being able to leave behind the responsibilities of being an IPCC author when back at home. Many times, I worked through weekends and evenings. My inbox continually groaned with hundreds of emails, as yet unread, discussing the latest developments in the crafting of the report. And each draft came with a hard deadline for completion that had to be met.

The closer we approached each deadline, the more it seemed there was for us to do. I finished work on the final draft in my office during a mammoth eighteen-hour stretch that only came to a halt when I was too tired to carry on. By that time, in the early hours in Britain, Nathan had arrived at his office in an Australian time zone half a day ahead, and I was able to hand over the text for him to apply the finishing touches. Thanks to those final efforts, our chapter was ready in time to be sent out for external reviews by scientific experts, to comment on its accuracy and comprehensiveness, and by government representatives to comment on its clarity and relevance.

Being a coordinating lead author of an IPCC report was hard graft. At times, it was frustrating having to deal with so many stages of preparation and review. But it was also rewarding. By comprehensively assessing the latest scientific research, we were providing governments with up-to-date and authoritative evidence on which they could base their domestic climate policies and engagement with international negotiations. And there was one aspect of climate change in particular that most governments were desperate to know more about.

The increasing toll of death and destruction from heatwaves, floods, droughts and storms was progressing at a startling and terrifying rate. The people affected wanted to know if they were just unlucky or if they were reaping the bitter harvest of years of industrial development based on fossil fuels. Policymakers were looking to IPCC to advise them on what they could say to the people affected.

At the time of the Fourth Assessment Report, there had been only one study that linked extreme weather directly to human-induced climate change. That was my paper: the one that linked the European heatwave of 2003 to greenhouse gas emissions.[6] This time, we had more of such studies to assess, reflecting an emerging scientific interest in trying to see whether heatwaves, floods, droughts and storms were linked to ongoing climate change.

The clearest evidence came from analyses of heatwaves. The Russian summer of 2010 brought unprecedented temperatures and, in its wake, thick smog from peat and forest fires that enveloped Moscow and other urban centres, killing more than 50,000 people. The record American temperatures of 2011 also caused a spate of wildfires as well as a record loss of crops in Texas, the centre of that heatwave. For both events, peer-reviewed papers had found that global warming had substantially increased their likelihood.[7]

Floods, droughts and storms were harder to attribute. There was one study of floods available that had analysed heavy rain in the UK in autumn 2000 that had flooded more than 10,000 homes and businesses.[8] Comparing thousands of climate model simulations that included the effects of greenhouse gas emissions and other human activities on climate with other model simulations that included only natural factors, this study concluded that anthropogenic climate change had significantly increased the risk of the flooding that occurred. But the uncertainties were large – with some simulations finding only a modest increase in risk and others a much more substantial one – and it hadn't yet been backed up by other research teams finding similar results for floods elsewhere. For droughts and hurricanes, there were no such studies then available. With research into attributing extreme weather events still in its infancy, for most types of meteorological event we had to be cautious. But for heatwaves, we were able to make an important statement.

We agreed the final version in our chapter room in Hobart: 'It is likely that human influence has substantially increased the probability of occurrence of heatwaves in some locations.'[9] Like a wolf in sheep's clothing, this sentence's cool IPCC language disguised a much more disruptive implication. No longer could it be claimed that all the victims, like the tens of thousands who perished during the heatwaves in Europe in 2003 and Russia in 2010, were simply unlucky. Their fate had been determined by human hands. Climate change was the culprit. The fingerprints proved it.

We broke up from our lead author meeting with a collective feeling of children being let out of school. Now, before we got back to work finalizing the report at our home institutes, we could look forward to rejoining families or taking a few days' holiday in an unfamiliar setting. Over the coming weekend, I had a day's hiking to enjoy with Nathan Bindoff and Thomas Stocker in a part of the Tasmanian countryside unaffected by the recent fires. But first, I had a talk to give at the university.

A Friday evening in midsummer was not normally a popular time for a public lecture. People would rather be enjoying the long evenings outdoors or be away on holiday. As we waited in a side room at the university for the event to begin, I was told not to expect the large hall where I was speaking to be particularly well filled. But when I stepped on stage to speak, I realized the place was packed. It looked like what I had to talk about was of great interest to an awful lot of people.

As I looked down at the sea of expectant faces in front of me, I felt a familiar sense of dread at being about to tell people things they might not want to hear. There could be climate deniers in the room, who would argue that I was deluded or corrupt for linking greenhouse gas emissions to an increased risk of heatwaves. And there could be others who were angry at the recent turn of events and who wanted me to account for why people hadn't had better warnings of the coming inferno.

The main focus of my talk was to present the evidence that burning fossil fuels had increased global temperatures, changed rainfall patterns and diminished snow and ice. I also dealt with changing weather and discussed recent research linking past heatwaves in Europe, Russia and America to human-induced climate change. With the current Australian summer still under way, we awaited a detailed investigation of the extent to which greenhouse gas emissions could be blamed for this most recent of heatwaves. But I stressed what was crystal clear: that worldwide, heatwaves and fires were on the rise. Tasmania had shared in this mounting devastation.

The questions after my talk showed people weren't interested in attacking the science of climate change. Instead, they wanted to know more about how their own experience fitted into the global picture and what could be done to prevent further damage. This was a new experience for me, meeting an audience who felt like this. At public talks in the past, the topic for discussion was usually seen by many as far off and hypothetical. Instead, for this group of people the issue was right here and frighteningly real. To serve these citizens better, we in the climate science community needed to understand more about how the weather was changing wherever those citizens lived. It was a global scientific challenge that had to be met soon, given the quickening pace of weather-related disasters worldwide.

The disaster unfolding in Tasmania, I told my audience, was a graphic example of the mounting risks of climate change, the state of knowledge on which had developed rapidly since the previous IPCC assessment of 2007. Notably, in a landmark paper published in 2008 in the *Proceedings of the National Academy of Sciences*, a team of scientists from the UK, Germany and the US had identified several key 'tipping elements' in the climate system.[10] These were aspects of climate which, if they were pushed too far, could suddenly and dramatically pass a tipping point and cause catastrophic and irreversible damage.

Sea ice in the Arctic was melting rapidly, but it could disappear completely in summer if warming continued to change the polar surface from white, heat-reflecting ice into dark, heat-absorbing ocean. Much of the Greenland Ice Sheet could also be doomed if temperatures passed a critical threshold beyond which melting of the periphery would lower the ice sheet's altitude and render it unable to survive. While its collapse would take three centuries or more, the resulting ice melt would raise global sea levels by between 2 and 7 metres, depending on how much of the ice cap survived. At the South Pole, the West Antarctic Ice Sheet could also collapse, undercut by melting from beneath. Once separated from its bedrock, a tipping point could be reached beyond which 5 metres of global sea level rise would be inevitable. And at the tropics, the Amazon rainforest, which generates much of its rainfall through local evaporation, could be condemned to inevitable collapse if drought and deforestation rendered it unable to sustain its natural climate. Such a regional tragedy would also have global implications. The great Amazon rainforest, the lungs of the earth, absorbs a substantial fraction of the carbon dioxide emitted into the air by human activities. The Amazon gone, global warming would accelerate even more.

Continued emissions could cause one or more of these tipping points to be passed in the next few decades. We were stumbling towards a cliff edge beyond which much of the earth would be uninhabitable and if current scientific understanding couldn't state precisely where that precipice lay, it told us that planetary catastrophe lay far too close for comfort. It all begged the question, as an audience member asked me, of why more, a lot more, wasn't being done to reduce emissions of greenhouse gases.

This question was the hardest of all for me to answer, given my powerlessness in the face of the ongoing and deeply frustrating political inertia in tackling climate change. The Copenhagen debacle was only three years in the past, and it wasn't yet clear whether renewed efforts in the climate negotiations would eventually bear fruit with an international agreement. Better warnings of fire weather coming people's way could help to some extent. But although such alerts could save lives, they couldn't necessarily prevent the destruction of valuable property and precious possessions. As recent events had shown, fires could be impossible to stop once started.

After the formal part of the evening was over, it was time to mingle with the crowd and hear about people's individual experiences. Everyone I met had felt the terrible oppressive heat of 4 January, smelt the acrid pall of smoke, and seen the clouds of ash fluttering down from black skies. But one man I talked to had experienced much worse. Like the Holmes family he came from Dunalley. And like them, his house had burnt down in seconds.

Amazingly, he seemed quite phlegmatic about the experience. He and his wife were alive, that was the main thing. His house could be rebuilt. What seemed to matter most was that he had saved one item money couldn't repurchase, the memory stick containing the now only surviving copy of his almost finished PhD thesis. I could identify with that. I'd slogged through a PhD myself, and the very idea of having to recreate the entire output from scratch made me shudder. Buoyed up by the immense relief of not having to redo three years of research, the chap from Dunalley seemed in remarkably good spirits.

Meeting these people helped me appreciate something of what it must take to live in a place like this. Like Nathan Bindoff, these inhabitants knew well enough that Australia was a continent that burned. They understood the ever-present danger of fire and knew there was a risk that one day their home might be destroyed. But they also felt strongly that their weather, brutal though it could be at times, should not be pushed into ever greater extremities. Learning more about the sequence of events in Dunalley helped me understand better the local implications of a global problem.

Fire risk is 'catastrophic' according to the Fire Danger Rating of the Tasmania Fire Service when the risk from extreme temperature, relative humidity, wind speed and vegetation dryness reaches its highest level.[11] That means temperatures over 40 degrees, winds over 100 kilometres per hour and less than 10 per cent humidity. Fires that burn on such days, the Tasmania Fire Service says, will be 'uncontrollable, unpredictable and fast-moving: flames will be higher than rooftops. Some people may die and be injured. Thousands of homes may be destroyed.' That is what was headed for Dunalley on 4 January. Yet the people who lived there didn't know what was coming their way until they were already in mortal danger.

The continental heart of Australia had been baking all summer. It was so hot that the country's Bureau of Meteorology had to invent a new colour for its weather maps, an incandescent purple chosen to represent forecast temperatures of over 52 degrees.[12] The island of Tasmania sits to the south of the continent, its temperatures usually tempered by winds from the cool Southern Ocean. But on that day, a vicious northerly was blowing, transporting overheated and over-parched air to seaside shores that were usually much more temperate.[13] To make matters worse, this was a foehn wind, in which the air was heated even more as it descended the southerly slopes of the hills surrounding the Tasman Peninsula. Like a giant hairdryer on steroids, this foehn wind fanned the flames, blowing burning embers from ridge top to ridge top and racing the fire front across the parched landscape.

Faced with this deadly combination of weather and climate – vicious northerly winds blowing straight from a continent baking in record-breaking heat – the homesteads of Dunalley didn't stand a chance. This was not climate change that could be adapted to. The Holmes family, whose picture had featured in newspapers around the globe, were going to rebuild their house as was the man I met in Hobart. But they knew that nothing they could put back would withstand another fire of such ferocity. Unless they were to leave Tasmania altogether, the only way the risk could be reduced, they knew, was for world leaders to take the one step they had so far signally failed to do and agree a rapid global reduction in emissions of greenhouse gases.

All the time, the planet's weather was becoming more and more extreme. The last year had seen a striking toll of weather disasters including: Hurricane Sandy, which devastated Haiti and slammed into the east coast of the USA, flooding the New York subway; Typhoon Pablo, which displaced 2 million people in the Philippines; and Typhoon Haikui, whose floods swamped roads and villages across large parts of mainland China.[14] Each year that went by saw the economic losses and humanitarian toll of meteorological catastrophes continue to mount.

Yet increasingly, a central plank of the argument made by climate deniers was that the world's weather was *not* changing. The Global Warming Policy Foundation, the well-funded climate-denial lobby group chaired by ex-Chancellor of the Exchequer Nigel Lawson, placed advocates like Lawson and writer Matt Ridley – whose respective books *An Appeal to Reason* and *The Rational Optimist* argued climate change was not an issue worth worrying about – on outlets like the BBC Radio 4 *Today* programme to argue that nothing out of the ordinary was happening with the weather.[15] Their claims ran counter to the mounting scientific evidence proving that greenhouse gas emissions were affecting heatwaves and floods, including in a series of annual reports on recent extreme weather events that I helped launch in 2012.[16] But still the misinformation continued: that climate and weather was just too complex to draw any conclusion other than that we didn't really know. As Lawson put it in a debate with one of the UK's most eminent meteorologists, Sir Brian Hoskins, on the BBC's *Today* programme, 'All I blame [climate scientists] for is pretending they know when they don't.'[17]

But climate scientists *did* know. And so too did the people I met who had been affected by the terrible fires that ripped through the Tasman Peninsula in January 2013. I owed it to them to make sure that governments had the best possible, most up-to-date scientific advice as they tried to build a new international consensus to reduce global emissions. In particular, a crucial new assessment was about to be released by the Intergovernmental Panel on Climate Change, the report that I had spent so much time working on. One week in Stockholm in September 2013 would decide what scientific ammunition nations would take into the next crucial stage of their negotiations.

10

False balance

I travelled to Stockholm for the approval session of the new Intergovernmental Panel on Climate Change (IPCC) report feeling a heavy weight of responsibility. The prospects for climate action depended hugely on what governments agreed over the next few days. My job was to convince them of the dominant role of human-induced climate change and support colleagues who would put forward new science showing there was little time left to avoid the worst effects of global heating. I didn't expect it to be straightforward. But failure was not something I wanted to contemplate. Without a shared agreement on the scientific facts, international negotiations to accept reductions in global emissions, due to culminate in Paris two years from now, were unlikely to succeed.

The day I went, I attended the formal adoption ceremony of the little girl of friends of mine. It was a happy occasion and I was sorry to miss the party afterwards as I had to set off for the station to travel to Sweden. But as I did so, I couldn't help thinking, given the mission I was embarking upon, what our country's climate might look like later in *her* life. Whatever sort of climatic disruptions she might face during the latter part of the twenty-first century – unprecedented floods, deadly heatwaves, devastating droughts – depended to a great extent on the greenhouse gases emitted while she was growing into young adulthood over the next two decades. One of the most important yet least appreciated aspects of climate change was that the long-term future was so closely tied to short-term actions. That was why delay in acting on climate change was so costly.

I did, however, travel in hope. By autumn 2013, the prospects for climate action were starting to look more promising than they had in the aftermath of the failure in Copenhagen in 2009. The annual meetings at which politicians haggled over the innumerable technical details that lay behind any global agreement to cut emissions – the Conference of the Parties to the United Nations Framework Convention on Climate Change (known as the COPs for short) – had been making progress. The 2011 COP in Durban had agreed that all countries, including the US, China and India, would take on future legally binding commitments on cutting emissions.[1] Exactly how deep those cuts would be was yet to be decided. But during the George W. Bush administration, the US had never ratified Kyoto, and China, India and other developing countries had had no obligations to reduce their emissions. Now, for the first time, as the science demanded, all countries were open to tackling the climate crisis together.

At the Doha COP the following year, developed countries agreed to carry on with emissions reductions under the Kyoto framework, which originally ended in 2012, for another eight years.[2] This was significant as without such continuity there would be no foundation on which to build a new global agreement that involved everyone. And ahead of the designated showdown in Paris in December 2015, there was one more piece of the puzzle that had to be slotted into place. That was a comprehensive new assessment from the IPCC, one that would underscore the reality, seriousness and urgency of the climate crisis.

Our meeting took place at a conference centre in a converted brewery that sat by the side of one of Stockholm's main waterways. Ahead of the government delegates arriving, we authors of the report began with a two-day preparatory meeting to finalize our headline messages and rehearse the arguments that supported our main conclusions. These preparations themselves were hectic and pressurized. Like swotting for a viva exam we had to pass first time – we knew we couldn't afford to arrive at the examination hall unsure of how to defend work on which so much depended.

There were three crucial aspects of the report that we desperately wanted to see approved. The first was a restatement of the vital sentence from the 2007 report, that warming of the climate system is unequivocal. After the baseless climate change denial unleashed during Climategate, the IPCC needed to restate the simple fact that evidence for global warming was beyond dispute. Next came the attribution statement we had decided upon in Hobart, that it was extremely likely that human influence was the dominant cause of warming since the mid twentieth century. To reinforce the message, I took a leading role in crafting an additional sentence that stated 'Human influence on the climate system is clear'. It was an important moment when everyone on the author team agreed that this extra statement would further underscore the message that – without doubt – climate was changing due to emissions of greenhouse gases. And finally, there was our conclusion about the urgency of reducing those emissions, based on new scientific understanding that global warming is directly related to the total sum of carbon dioxide emitted so far.

As well as fine-tuning our statements and rehearsing our arguments for their defence, a number of us practised the presentations we would give to governments on different aspects of the report. It was nerve-wracking delivering mine on the causes of climate change to my peers. A lot depended on me to help steer a successful outcome on this crucial aspect of the report and my presentation would be a key part of that. But my colleagues had some useful feedback and if my nerves were tempted to get the better of me, I should remember that our conclusions had already survived the critical scrutiny of many of the world's foremost scientific experts. After two days of thorough preparation, I felt ready to face the critical scrutiny of the world's governments.

When the approval meeting began on the morning of 23 September 2013, I felt well prepared but apprehensive.[3] The stage was set for what could be a momentous few days. I had a grandstand view from the section reserved for authors of the report at one side of the high-ceilinged room. In front of me was the raised platform from which Thomas Stocker would chair the meeting alongside his co-chair Qin Dahe and to my right were the long lines of tables where the various delegations sat. As usual, countries were arranged in alphabetical order, with Australia and Austria near the front and the UK and the US at the back. Above me was a balcony on which the interpreters' booths had been placed, enough for all the six UN languages to be simultaneously translated, one into another.

With everyone in place, I felt a keen sense of anticipation at being about to take part in history being made. In due course I would join Thomas on the raised platform and defend our conclusions on the causes of climate change. In the meantime, I reflected on the need for this week to mark a decisive turning point in the climate crisis. Warming is unequivocal. Human influence on the climate system is clear. Once that was accepted by everybody here, the nonsense of climate change denial would be dead and buried as a global political force. Policymakers could get on with hammering out the terms of an international deal to avoid climatic catastrophe. The future for everyone would look a little bit brighter.

Proceedings kicked off with a brief press conference to announce the start of this landmark meeting. Short speeches were made by dignitaries from the United Nations, the Swedish environment minister – 'I just wish that those who question climate science would follow the scientific rules as well', she said – and Rajendra Pachauri, the Indian scientist who acted as Chair of the Intergovernmental Panel on Climate Change.[4] It was Pachauri's role to lead a committee that oversaw the work of the IPCC including our part of the process, the working group assessing the physical science basis of climate change.[5] He would stay for the rest of the meeting but he was mainly expected to stay silent and observe.

The journalists were asked to leave and after Thomas had made a short presentation about the structure of the report, the first sentence for approval was placed up on the giant screen at the front of the hall. From here on there was no turning back until every single word of our Summary for Policymakers (the SPM for short) had been approved by every government present. We had to be prepared for long nights, lengthy debate and possible questions about every last detail of the underlying 1,400 page report.[6] And we had to finish in time for the press conference at 10 a.m. on Friday morning at which our results would be presented. Ahead of us was a daunting prospect. But I was eager to get going and see how governments reacted to what we had to present.

We started with the introduction to the SPM, a short section outlining the structure of the document and describing the terms used to express the degree of certainty in the key findings of the report. The first sentence was a simple statement that the report considered evidence of past and future climate change from observations, theoretical studies and climate models. The first intervention came from the Saudi delegate. The report needed to strengthen IPCC as an authoritative body, he said. While we have observations of the past, he continued, the future is only available from simulations that have a lot of assumptions and therefore the sentence should not include the words 'past' and 'future' together. After a lengthy discussion, during which many nations spoke up to support the authors' inclusion of these two words but Saudi Arabia continued to argue against, the sentence was eventually approved with 'past and future' removed. Forty minutes had elapsed and, already, the oil-rich nation had given a signal that later in the meeting they could be resistant to some of the main messages of the report.

I found this more alarming than the snail-like pace at which we progressed through subsequent sentences during the rest of the first morning's session. I was familiar with the slow start of such meetings from my experiences approving the previous IPCC report six years before, but it came as a shock to colleagues new to the process. It was all right, I knew, as long as the pace picked up soon and individual countries weren't allowed to derail later discussions with unreasonable objections. And if agreeing the document word by word felt like a tortuously laborious process, we scientists should remember the prize at stake. With the scientific evidence formally accepted by every country, responsible governments could not derail political negotiations later by using apparent disagreements about the science as an excuse for delaying action on addressing climate change.

Agreeing the seemingly anodyne words of the introduction took up most of the day and progress on Tuesday was little better. Many countries were unhappy that we had not provided the amount of warming since pre-industrial times. In Paris two years from now, their governments wanted to set targets above pre-industrial levels that global temperatures should not exceed: many wanted a target of 2 degrees Celsius while some low-lying island states threatened by sea level rise wanted a stricter target of 1.5 degrees. In Stockholm, they wanted us to say how much warming there had already been, relative to those pre-industrial times, so they would know how much leeway there was before their targets would be breached. Unfortunately, they had set us an ill-posed question.

The Industrial Revolution in Europe and the United States took place from about 1760 to the early decades of the nineteenth century, making the pre-industrial period any time before the mid eighteenth century.[7] Global average temperatures had been very stable for millennia during those pre-industrial times, but they did vary slightly – by a few tenths of a degree – largely due to volcanic eruptions and subtle changes in solar output. Furthermore we couldn't measure these changes very precisely as we didn't have instrumental records of global temperature before the mid nineteenth century and we had to rely on proxy records like tree rings. In summary, we couldn't give the policymakers exactly what they wanted, a precise measure of temperature changes since pre-industrial times.

With the meeting unable to resolve the issue in the main room, animated discussions continued in the lobby between authors from the observation chapters and representatives of the delegations. It seemed extraordinary to me that policymakers had not realized the problem sooner. But now that they *had* realized, many delegates were pushing us to fix this problem by coming up with a way of assessing the progress of global warming against the temperature targets they wanted to set in Paris. Given the importance of the issue and the difficulties involved, colleagues with expertise in estimating temperatures during past centuries were looking very worried indeed.[8]

Another sticking point was how to characterize the rate of warming in recent years. The apparently slightly lower rate of warming at the surface over the last fifteen years compared to over the last sixty years, which we had quoted in the report, had become a cause célèbre for climate deniers. They had used it to support claims that global warming had slowed down or stopped altogether, even though it was well established that surface warming rates under a substantial long-term warming trend could vary temporarily. I had demonstrated this fact in my very first climate science presentation at an international conference – the session in San Francisco in 1996 when Ben Santer had been confronted by Patrick J. Michaels. We had reiterated this fact in our report of 2013. A better indicator of global heating, we explained, was how much heat was being taken up by the world's oceans.

The latest measurements of sub-surface water temperatures showed the oceans were heating steadily, absorbing over 90 per cent of the extra heat trapped in the climate system from greenhouse gas emissions. This showed that global heating had not stopped or slowed down but was continuing unabated. A vivid demonstration of this fact came quickly in the years after 2013: the following year 2014 broke previous records for surface temperatures and was followed by a sequence of extremely hot years, every single one of which was hotter than any year before 2014. By 2020, the surface warming trend over the previous fifteen years had risen to values higher than the warming rate over the last sixty.[9] If, in 2013, the climate deniers had been crowing about a global warming slowdown, by 2020 they were definitely not talking about what by the same logic they should be calling a global warming surge. And all along, the oceans told their own terrifying story – a remorseless rise of global heating that showed no sign whatsoever of slowing down.

The information about ocean heating was contained later in the SPM. But for some delegations, quoting a number for the surface temperature trend over the last fifteen years was problematic. They feared it would be cited by countries keen to stall negotiations on reducing greenhouse gas emissions. To us authors of the report, we couldn't hide the facts, even if some delegates might regard them as politically inconvenient. Politicians needed to look at all the evidence we had to present, not misuse one sentence taken out of context.

By lunchtime on the second day, with little progress being made on both of these issues, the atmosphere was becoming fractious. Thomas appealed to delegates to come along with the proposals of the scientists. We are making almost no progress, he told them. This prompted an angry response from the head of the Saudi delegation who accused Thomas of seeking to misrepresent governments' views by not calling sufficiently on his country.

Over the lunch break I gave my talk on the causes of climate change. Delegates listened intently and there was time for a couple of questions of clarification at the end. Getting back to the basics of our scientific evidence seemed welcome to everyone. I finished with the headline statement I'd agreed with colleagues in our preparatory meeting: 'Human influence on the climate system is clear'. Governments would consider the text formally later in the meeting. But so far it looked like there were no objections and the Saudis had stayed quiet – for now.

In the afternoon, progress through the document remained painfully slow. To catch up time, an additional session was scheduled for the evening. The issues about warming rates remained unresolved and on the next day, Wednesday, were still being debated in side rooms around the conference. Still, each new sentence needed to be debated and then approved. For much of the time I was biding my time, worrying about the slow progress and fretting about when I would finally get my chance to join Thomas on the podium and defend our part of the assessment.

Finally, an agreement was reached as to how to deal with temperature changes since pre-industrial times. The pragmatic solution was to express warming relative to the average temperature over 1850 to 1900.[10] Without widespread thermometer readings before the late nineteenth century, it was as near to an accurate measure of temperature change above pre-industrial conditions as it was possible to get. Estimates of temperatures from tree rings showed that choosing an earlier period made relatively little difference. And after colleagues calculated fifteen-year temperature trends starting at a range of different dates from 1995 onwards, delegates agreed to keep the fifteen-year trend from 1998 in the report, provided the other calculations were included in a footnote. With these issues resolved, the mood among the author team lightened as optimism rose that we might be able to finish in time for the press conference on Friday morning.

Eventually, at 10 p.m. on Wednesday evening, we came to our crucial headline statement about the basic facts of global warming. Now that we had reached this pivotal moment in the week's crucial deliberations, all of my colleagues were looking up at the words that appeared on the giant screens before us and wondering what would happen next. 'Warming of the climate system is unequivocal', the two-sentence summary stated, 'and since the 1950s, many of the observed changes are unprecedented over decades to millennia. The atmosphere and ocean have warmed, the amounts of snow and ice have diminished, sea level has risen, and the concentrations of greenhouse gases have increased.' We all shared a strong conviction that these facts of the matter needed to be recognized by every government present, including those who had been most resistant to such facts in the past.

Unsurprisingly, the head of the Saudi Arabia delegation was the first to respond. 'This is obviously your reading of the message,' he said. 'It's an extremely dire situation out there. We are using such big words that need qualifiers. This is generic wording that if you were to leave it there you would scare everyone. It is not a true reflection of the true science.'

I was familiar with these sorts of arguments as they had been levelled at us frequently by critics like Nigel Lawson and Matt Ridley. Claiming to be the voices of reason and rationality, they might concede that some warming was taking place while insisting that climate scientists grossly exaggerated the possible threats to try to scare everybody into expensive and uncalled-for actions. Like the Saudi delegate, they claimed that we were being alarmist when such alarmism wasn't supported by the science.[11]

Yet the science did support raising the alarm. The risks of climate change were becoming clearer all the time. The financial and human toll from weather disasters was rising substantially.[12] The earth's ecosystems were being damaged beyond repair. The West Antarctic Ice Sheet was perilously close to collapse risking many metres of sea level rise. If anything, for risks like the West Antarctic Ice Sheet collapse or the possible shutting down of the ocean conveyor in the North Atlantic, the IPCC could *underplay* the seriousness.[13] For some risks, whose probability was relatively low but whose impact was unimaginably severe, restrained IPCC language of low likelihoods could disguise the full implications of a risk that was actually quite substantial. Even so, no reasonable or rational policymaker could read our IPCC report and not accept the overwhelming evidence we had presented that climate change posed alarming risks. They were risks that needed to be mitigated before it was too late.

Nevertheless, the Saudi delegate continued to insist that we were being alarmist. 'We have to remove it altogether', he said referring to our summary sentence, 'or qualify the words unequivocal or unprecedented.' With such a clear statement of resistance, I waited apprehensively to see whether other countries would lend their support.

Around the room, a host of nations had placed their nameplates in the vertical to indicate they wanted to speak. Thomas called each of them in turn. One by one, they provided unqualified support for the unequivocal and unprecedented statement. 'We think this is a very good headline,' the German delegate said. The Australian representative agreed. 'It is a good reflection of the statements we've had on observations over the last three days,' he said. 'It is time we move on.'

Still the Saudis were not prepared to back down. 'I still think it is an inflammatory statement that is not warranted,' the leader of their delegation declared. I wondered whether they were waiting for an ally to speak up in their favour. But all the delegations who spoke next failed to back him. Instead, countries lined up – Spain, New Zealand, Slovenia, the UK – to express their endorsement of the statement as it stood. Back Thomas came to Saudi Arabia but again he was rebuffed. 'This sounds to me a very emotional statement,' the country's representative said.

At this critical juncture, the IPCC Chair, Rajendra Pachauri, made an unexpected intervention. Surprisingly, he chose to quote statements not from the report we were considering now, but from the IPCC's previous report that I had seen approved in Paris in 2007. 'Warming of the climate system is unequivocal', the Indian scientist read out, 'as is now evident from observations of increases in global average air and ocean temperatures, widespread melting of snow and ice, and rising global average sea level.'[14] He was making the point that the IPCC – including the Saudi delegation – had already agreed six years ago that warming was unequivocal. With even more evidence of a warming climate now, how could we not endorse a similar statement this time?

Finally Saudi Arabia backed down. It was good news, not just that the unequivocal statement had been approved, but also that no other country had tried to block it. The Middle Eastern nation had been left isolated in the face of overwhelming evidence against them. They had no alternative but to let us all move on to the rest of the document.

After these moments of high tension, Thomas pressed further on into the night with a section of the document about the ability of climate models to predict climate change. Despite the repeated claims of climate deniers that climate models were not up to the job, predictions documented in past IPCC reports stretching back to 1990 had all succeeded in capturing the subsequent warming.[15] As our summary statement said, 'models reproduce observed continental-scale surface temperature patterns and trends over many decades'.[16] It took several hours, but eventually by 2.30 a.m. all the sentences of this section had been accepted. Tomorrow morning, we could start to consider my section of the document, the one on the causes of climate change.

After snatching a few hours' sleep, at 8 a.m. that same morning I was back in my place for the most testing day of all. The next few hours weren't only a test of our work over the last couple of years preparing our report. They were the culmination of over two decades of research into the causes of climate change and the likely nature of future warming. Success or failure would have huge repercussions for the crunch summit in Paris just two years from now. It all depended on the sentences we proposed to delegates being approved as accurate summaries of the latest scientific understanding.

To begin with, our sentences on the causes of climate change were accepted quickly. They described the role of human influence on atmospheric and oceanic temperatures, on rainfall patterns, on loss of Arctic sea ice and on sea level rise. Nobody wanted to question them when the evidence we had compiled was so convincing. The statement that we had crafted in Hobart in the immediate aftermath of Tasmania's terrible destructive fires – 'human influence has more than doubled the probability of occurrence of heatwaves in some locations' – also passed unopposed.[17]

The careful way we had justified our headline conclusions seemed to be paying off. The table we had compiled for our chapter, with thirty-three entries that itemized in great detail the scientific evidence supporting each of our headline statements, meant that delegates had in front of them a clear summary of how we had come to our conclusions. The evidence for human influence on climate lay everywhere you looked, the table made clear, on land and sea, on the ice and in the air.

Thomas, sticking to the usual procedure in IPCC approval sessions, had left the most important sentences in our section of the SPM, the ones that provided the highest level of headline conclusions, to last. 'Human influence has been detected in warming of the atmosphere and the ocean', our summary of summaries read, 'in changes in the global water cycle, in reductions in snow and ice, in global mean sea level rise, and in changes in some climate extremes. It is extremely likely that human influence has been the dominant cause of the observed warming since the mid-20th century.'

Given the rapid progress so far, I thought this too would pass unopposed. But when these words were placed on the giant screen for everyone to see, a member of the UK delegation raised their flag to object. She told us that the statement needed to recognize that since the previous report of 2007 there had been a growth in evidence of human influence on climate. A host of other nations raised their flags to support the British intervention. Slovenia, Switzerland, Canada, Fiji, Saint Lucia and Germany all spoke, proposing various amendments to our text to make this point. Saudi Arabia spoke too, but they wanted to object to any such change.[18]

Sitting on the podium next to Thomas and my co-lead of this section of the report, Nathan Bindoff, I was happy to acknowledge that the evidence had indeed strengthened. But wordsmithing in the main hall wasn't producing an outcome that I or Nathan could agree to. The problem was, as I explained to the delegates in front of me, that it wasn't possible to directly compare the attribution statement from last time – that 'most of' the warming since the mid twentieth century was 'very likely' due to the increase in greenhouse gas concentrations – with our current formulation that human influence was 'extremely likely' the 'dominant cause'. If it seemed like nit-picking to some delegations, Nathan and I were sticking to our guns. Every word counted. Every word had to be supported by the evidence we'd laid out in detail in our chapter of the report. If we didn't do that, we would have failed to provide the best scientific advice to governments and opened up the report to attack by climate deniers.

With time ticking on into the late afternoon and with reams of the SPM still to approve, Thomas had no alternative but to bracket the two offending sentences and instruct Nathan and I to leave the room and gather together with interested parties to find a form of words that everyone could agree to. Nathan and I set off to a side room behind the main hall. Nathan put his laptop on a high table and we waited while a large huddle of government representatives formed around us.

For a while, we spent time going over the arguments we'd already made in the main hall. We couldn't agree to an ill-posed comparison between the two headline statements from the two reports. How to avoid this, while also acknowledging that the evidence for human-induced climate change had indeed grown, did not seem obvious. It was a puzzle we had not been able to solve earlier, under the glare of hundreds of delegates. But suddenly, as the pressure mounted for us to come up with something satisfactory, else risk either downplaying the case for human-induced climate change or else agreeing to something that could be undermined by rigorous scrutiny later, we realized there was, after all, a simple solution.

Compared with six years ago, we now had much more evidence of the effects of human-induced emissions, not just on rising temperatures but on changing rainfall patterns, melting snow and ice, mounting sea levels and climate extremes. But rather than try to compare two incompatible sentences about the amount of global warming attributed to human factors, what we *could* do was add an extra sentence to follow the one describing the detection of human influence in many aspects of climate. As everyone in our huddle looked on, Nathan typed it into his laptop at the appropriate place in the document. 'This evidence for human influence has grown since AR4,' it read, AR4 being the Fourth Assessment Report. It was a simple, well-justified form of words that expressed the fact that there was now more evidence of change from across the earth's climate than ever before. Happily, it was one that everyone could sign up to.

Except, perhaps, Saudi Arabia. 'How could we justify such a claim?' their delegate asked Nathan and me. I explained to him, once more, that we had a wealth of new evidence to draw on. What's more, I again set out that it was fully itemized in thirty-three detailed entries in the extensive table to be found in our chapter.[19] Hearing this, he left the group saying he would have to check with his head of delegation. He returned quickly, thankfully to signal his assent. Walking back to the hall with consensus achieved, Nathan asked him if he was happy. 'No', he laughed, 'I'm in agreement,' and clapped Nathan on the back.

Our revised attribution statement was put to all of the delegates later that evening. Up on the podium once more, Thomas asked me to explain the reasoning behind it. I told them there was now even more evidence of the effects of human-induced emissions on climate than there was at the time of the Fourth Assessment Report six years ago. This evidence was set out in detail in our chapter including an extensive table itemizing the many aspects of climate change which had been detected on land, on sea, on ice and in the air. We waited to see if there were any more objections. None came. The statement had been approved.

It was a relief that such a statement had been passed. And later that night, another important landmark was reached when the sentence that I had taken a leading hand in developing, 'Human influence on the climate system is clear' was also passed unopposed.

By this stage, I was exhausted but delighted. Climategate now firmly behind us, its claims of fraudulent misdirection by the scientific community now definitively disproved, governments had unanimously accepted the reality of human-induced climate change. Later I would be able to relax and reflect on what we had achieved. But for now, in the early hours of Friday and with only a few hours left before the press arrived to discover what we had concluded, there was still one more crucial section of the report to be agreed. And without it, none of what had gone before would have had much relevance. It was the section concerning the science behind the urgency of tackling climate change, the justification for why humanity couldn't wait any longer before reducing its emissions of greenhouse gases.

Swiss scientist, Reto Knutti, took the lead in defending this part of the report. Reto had studied for his doctorate under Thomas and like his former supervisor spoke Swiss German, a dialect only understood in his native country, which meant they could swap notes privately in Swiss German while publicly addressing each other and the government delegates in English. It was now their job to defend the statement displayed on the giant screen at the front of the hall for all to see: 'Cumulative total emissions of carbon dioxide and global mean surface temperature response are approximately linearly related'.

Delegates and scientists alike knew this rather dry-sounding result held enormous implications. A direct relationship between emissions and global warming meant that a failure to reduce emissions in the next few years would require much greater efforts later to meet the temperature targets being negotiated by governments, efforts that could soon become impossible if emissions carried on unchecked for much longer. If you wanted to keep temperature rise to a certain level, there was only so much carbon you were allowed to burn.

As soon as Thomas opened the IPCC's new statement up for debate, China, Saudi Arabia and then India questioned the validity of this startling new claim. How could it be, they demanded to know, that there was such a simple relationship between total emissions and global temperature rise? Thomas agreed that this result was a surprise but he also insisted that discovering it was a major scientific achievement. Thomas asked Reto Knutti to explain.

With higher emissions, Reto told the delegates, the ocean takes up less carbon from the atmosphere, increasing the fraction remaining in the air. This adds to the rate of warming at the surface. But with higher emissions, the ocean also takes up more heat, which reduces the rate of warming at the surface. The two effects cancel each other out in such a way that for every additional unit of carbon dioxide emitted into the air, there is an additional increment in global temperature increase. Double the carbon dioxide emitted into the atmosphere and the global temperature rise doubles, halve the amount emitted and the temperature rise halves. Nature really is that simple and there is a wealth of scientific evidence to prove it.[20]

Despite Reto's explanation, backed up by the thoroughly reviewed assessment provided in the body of the report, some government representatives continued to question whether this linear relationship really was correct. In recent years, they said, temperature had not gone up linearly with rises in carbon dioxide. But, Reto explained in response, it was crucial not to confuse the short-term, year-to-year variations of global temperature due to natural variability with the steady long-term trend in human-induced warming. It was over these longer timescales that total emissions were related to temperature rise. And it was the longer-term rise that mattered for temperature targets.

With time ticking on towards two in the morning, the scientific author team were very tired. But most of us were still wide awake, weary eyes glued to the unfolding drama. Our scientific message, unwelcome though it was, had to be recognized and acted upon. But it seemed like some of the very countries we wanted to help with our advice, including China and India who stood to suffer devastating floods and debilitating food shortages if emissions carried on as they were, seemed reluctant to heed it. Thomas spoke out to remind everyone that the IPCC authors of the report had been asked to comprehensively assess the science, which we had done. If we had not done so, he told them, you would not have been happy. The Irish delegation intervened to say that this was a beautiful and elegant finding and that they were surprised that it was taking so long to adopt it. China appeared to soften, commenting that the language had improved over previous revisions of the report. Finally, after over an hour of debate and with nobody able to convincingly argue against the strength of evidence we had provided, the statement was approved.

We still had one more crucial statement to agree and it was potentially the most contentious of all. This one put numbers to the linear relationship. It was not possible to know the exact factor by which emissions caused warming, which meant that temperature targets had to be related to cumulative emissions probabilistically. But the research showed that if you wanted to keep warming to below 2 degrees Celsius with a probability of at least 66 per cent, cumulative emissions over all time had to be kept to less than 2,900 gigatonnes of carbon dioxide.[21] By 2011, 1,900 gigatonnes, two thirds of this maximum total, had already been burned through. With carbon dioxide being emitted at unprecedented rates, there was little time left to have a decent chance of avoiding breaching the 2 degree target. Keeping temperatures to below 1.5 degrees Celsius would be even harder.

China was the first country to respond to this sobering conclusion. They wanted clarification on the numbers – why the target of 2,900 gigatonnes was a single value and not a range of values. Reto explained that the figure was associated with a probability – greater than 66 per cent – of staying below 2 degrees Celsius warming. By quoting the numbers this way, the analysis had already taken account of the range of possibilities. If the world emitted more than that, the probability of staying below warming of 2 degrees would be lower. The Japanese delegate wanted to know why 66 per cent had been chosen. Reto said that this was because it was a judgement of what the world believes is justified. A higher probability of avoiding warming of 2 degrees would imply a lower maximum total of emissions to achieve it. Hearing this, the leader of the US delegation raised his flag to make an intervention. It was a striking one.

'The IPCC is supposed to be policy neutral,' the American told the assembly. 'A single number would pre-empt the political discussions needed to decide the risk humanity was prepared to bear. The IPCC's job was to inform those discussions by linking emissions to temperature targets, not to tell negotiators what was an acceptable probability for avoiding a particular temperature target.'

Chastened by this pointed intervention, Thomas had no alternative but to instruct interested parties to leave the room and decide upon a better wording outside while he carried on working through other statements. He knew that IPCC assessment had to be policy relevant but not policy prescriptive. The American was right. With time rapidly running out before the press conference to release the report, Thomas had to hope Reto could figure something out quickly, an alternative that wouldn't imply scientists rather than policymakers had decided what risk the world should be prepared to bear.

In the foyer, a large group gathered round one of the high coffee tables. The head of the American delegation made a short speech explaining that IPCC attaching a single number to a target would be policy prescriptive because it would imply that a 66 per cent probability of breaching warming of 2 degrees Celsius was an acceptable level of risk. That was a matter for people and their governments to decide, not scientists. Instead, quoting a range of numbers representing different probabilities of staying below 2 degrees warming would be more appropriate. Then policymakers could look at different levels of risk and decide on the implications for themselves.

Fortunately, Reto knew how to calculate the additional numbers. Ahead of time, in anticipation of possible emergencies like this, he had asked one of his team of postdoctoral researchers to keep watch at his computer through the night, ready to make any additional calculations that might be needed. As the debate unfolded, Reto sent out an email to the deserted Zürich lab telling his colleague that his attention could soon be required, depending on the decision made in the next few minutes.

After some discussion, the group gathered around the foyer coffee table agreed that other emissions totals should be quoted corresponding to 33 per cent and 50 per cent probabilities of staying below 2 degrees Celsius of warming. If we could get the numbers calculated in the next hour or so they could be included in the report. If not, the American delegate was adamant that it would be too late. The report had to be signed off overnight as delegates had booked flights home in the morning. The implication was clear: if Reto's postdoc researcher wasn't able to make the calculation soon, the report would lose one of its most significant conclusions.

At 3.30 a.m. in the main hall, Thomas presented the new sentence with gaps where the new numbers would be inserted as and when we had them from Switzerland. Thankfully, and thanks to the efficiency of Reto's postdoc, they arrived surprisingly quickly. These are 'more informative and less cherry-picky numbers,' said the American delegate when he saw them appear on the screens at the front of the hall. 'It wouldn't win a Pulitzer Prize,' said Thomas, reflecting the slightly tortured nature of the English we'd arrived at in the extra information. Crucially though, it provided guidance on allowable carbon dioxide emissions without breaking the IPCC's remit to avoid being policy prescriptive. Consensus reached, Thomas moved on.

It took another hour and a half for the rest of the document to be approved. On the home stretch but running on adrenaline, Thomas became ever more loquacious. As IPCC Chair Rajendra Pachauri put it in a light-hearted intervention, he was getting more and more eloquent as he was getting more and more tired. At last, to general applause, the final sentence of the 13,000-word document was gavelled down at 5 a.m.

My colleagues and I walked back to our hotels dog-tired but very happy. It was still dark, although with dawn not far off we could only snatch a couple of hours' sleep before we would have to return for the press conference. But we had done what we set out to do. Governments had accepted our conclusions. Now they would have to take them on board during the negotiations leading up to the decisive showdown in Paris in 2015. Whatever happened, nobody could say the politicians didn't have the evidence they needed to act. We could be proud of what we'd achieved in Stockholm.

I was back at the congress centre for the press conference at ten. Like in Paris six years before, the room was now occupied by journalists rather than government representatives. But unlike in Paris, when I had witnessed what was probably the best attended media gathering climate science has ever seen, this time the room was sparsely populated. British journalists dominated the questions to the panel which included Thomas, Rajendra Pachauri and dignitaries from the United Nations. Afterwards, I gave an interview for the BBC *Newsnight* programme in the foyer where the drama on carbon emissions with the American head of delegation had taken place just a few hours earlier. Then I went outside with BBC environment correspondent Matt McGrath to be interviewed for Radio 5 Live. It was a straightforward set of questions about the findings of the report, which I answered standing near to a large melting block of ice that had been left there by a group of activists from Greenpeace. After the interview he told me that the BBC wanted me for another interview with the Radio 4 programme *World at One*.

That was a surprise. It was also a surprise when the BBC reporter handed me a set of headphones and his microphone to hold and walked away from me as though he wanted nothing more to do with what was about to happen. Then came a worse surprise. Through my headphones I heard the radio producer in London telling me that they had already interviewed Professor Robert Merlin Carter for their item on our IPCC report. This was ridiculous. On the day of the most important scientific announcement on climate change for six years, the first news and current affairs programme to broadcast after its release was featuring one of the world's most prominent climate deniers. I had come across Carter when he was my opposite witness during the High Court trial into *An Inconvenient Truth*. He spouted nonsense then and he would have spouted nonsense now. But not having been forewarned about all this, it felt very much like an ambush, an attempt to catch me off guard in the hope that, tired and sleep-deprived, I'd say something unhelpful about our findings.

The first question from presenter Shaun Ley confirmed my fears that this was going to be a hostile interview. 'Aren't the conclusions released today just the result of government-influenced haggling?' he asked me.[22] Further questions continued in the same vein. 'Wasn't warming now proceeding at a slower rate than previously thought and shouldn't that inject a note of caution?' 'Weren't the previous IPCC conclusions of six years ago criticized?' 'Hadn't the IPCC made errors in the past?' Faced with this critical interrogation, I determined to tell Radio 4 listeners what they actually needed to know rather than amplify irrelevant and unwarranted criticisms that had come presumably from Carter.

'The process is extraordinarily scientific,' I replied in answer to the first question. 'We've gone into the science in extraordinary detail,' I told the listeners, 'We have got this statement, very clearly expressed, that human influence on the climate system is clear, and that the confidence is greater than 95 per cent for the dominant role of human influence on climate.' Recent data did not inject a note of caution into the debate but instead strengthened our conclusions. 'In fact, in the last fifteen years, we've presented evidence that the oceans have warmed', I said, 'that the snow and ice has melted, the sea level has continued to rise, and extremes have continued to become more intense and more frequent.' And as to errors and previous conclusions being criticized, I stressed how rigorous this new assessment was. It was based on peer-reviewed literature, on responding to over 50,000 review comments on drafts of the report and on scientists presenting the detailed scientific evidence line by line to governments here in Stockholm.

Listening to the broadcast later was infuriating. The headline introduction to the lead story on that day's bulletin did state that our long-awaited report had concluded that it was 95 per cent certain that human activity was the main cause of global warming. But this was immediately followed, not by me or another IPCC scientist, but by a quote from Carter saying that no government tries to stop an earthquake or a volcanic eruption and similarly no sensible government should dream of trying to stop climate change. And after a brief sound bite from the UK's Secretary of State for Energy and Climate Change, Ed Davey, saying that the IPCC report was 'probably the most robust, rigorous, most peer-reviewed piece of science in human history', Carter was given ample time to expand on his view that the IPCC's conclusion about human activity was 'hocus-pocus science' and that the only response needed to climate change was to adapt better to natural disasters. My interview followed and was given roughly the same time as Carter's. It was a classic example of 'false balance'.

In 2011, an independent review into the impartiality and accuracy of the BBC's coverage of science conducted by geneticist Steve Jones had warned against false balance.[23] His report concluded that the BBC had been prey to falling into this trap in the past when reporting on contentious issues including climate change, and in future it should avoid giving equal weight to consensus and minority views. Two years on, it seemed that this advice was not yet being heeded. On the day of our landmark report with its far-reaching implications, more damage had been done to the public understanding of this most pressing issue of our times.

I could only hope the public would see through such false balance. On them, after all, did everything depend – on their individual actions and on their democratic choices and political actions. Freed of the heavy demands IPCC authorship requires, I would soon have more time to devote to my own research and the outreach activities I now wished to pursue. The political process of climate action was gathering pace, as national leaders contemplated attending the make-or-break gathering in Paris in December 2015. And travelling to Dublin in September 2017, I would discover what citizens could do when countries trusted them to consider the evidence for themselves and let them recommend how their governments should respond. This, I was to discover, would be a crucial instrument in the battle against climate change.

11

Citizen power

I had been to Ireland only once before. It was many years ago, when I was a student, and I'd always remembered how friendly the people were. But now, here in Malahide, a coastal town 16 kilometres north of Dublin, that seemed a distant memory. I had come to participate in one of the most fascinating exercises in deliberative democracy anywhere in the world. Members of the Citizens' Assembly of Ireland were meeting to consider how to make their country a leader in tackling climate change. I was one of the experts they had asked to brief them.[1]

I looked through the open doorway to the crowded bar of the Grand Hotel, and suddenly felt reluctant to enter the fray. Instead, I walked through the hotel's front door and took a turn outside. I felt nervous about introducing myself to these people I didn't know when they were all apparently getting on like a house on fire. Even so, joining this pre-dinner gathering was the first task of my weekend's work. After a couple of circuits of the car park, I took a second pass at entering the busy room. There was a plush carpet, comfortable armchairs and a wood-panelled bar in front of which a bustle of people stood chatting and ordering in the next round of drinks. Feeling like a stranger in a group of friends, I looked around for a familiar face.

I needn't have worried. I had been spotted by Anna Davies, professor of geography at Trinity College Dublin, who was part of the Expert Advisory Group that had invited me here. In no time at all, I was installed at a table in the middle of the cosy room with a drink in front of me and had been joined by some of Anna's colleagues, including Áine Ryall, a senior lecturer in environmental law at University College Cork, and John Garry, an expert in deliberative democracy. While we waited to go into dinner, they told me about their role in the forthcoming deliberations.

The role of the Expert Advisory Group was to help construct a suitable work programme for the Assembly's deliberations on climate change and to help choose the group of experts who, like me, would be called to give evidence. Other members of the group included Diarmuid Torney, lecturer in international relations at Dublin City University, Peter Thorne, professor of physical geography at Maynooth University, and Margaret Desmond, research specialist at the Environmental Protection Agency Ireland. But extraordinarily, despite their wide-ranging knowledge and insights, they were *not* the ones who were going to be advising their government. That job was down to the other people here, the members of the Irish public, laughing and chatting around me.

The formal part of the meeting began the next day, Saturday 30 September 2017. The first part was private and open only to Assembly members and the Expert Advisory Group, although guest speakers like me were also allowed to attend. It consisted of a short discussion of the weekend's business led by the Assembly's chairperson, the Honourable Mary Laffoy. After this, the meeting was opened up to the public and journalists and the internet live stream began. It was a fascinating process which I felt privileged to be part of.

The Citizens' Assembly was set up in 2016 by the Oireachtas – the Irish Parliament – to consider a limited number of contentious issues and to provide recommendations.[2] The Fine Gael–Independent government had proposed considering four issues: the Eighth Amendment of the Constitution (the controversial provision that

banned abortion unless the life of the pregnant woman was at risk),
the challenges and opportunities of an ageing population, fixed-term
parliaments, and the manner in which referenda are held. A fifth issue
was added by the Green Party when the proposal was debated in
the Dáil Éireann (the lower house), which was how to make Ireland
a leader in tackling climate change. The Assembly consisted of an
appointed chair and ninety-nine randomly selected individuals who
were drawn to provide a representative cross-section of Irish society. It
first met in September 2016. One year on, after considering the Eighth
Amendment and an ageing population, it began its deliberations on
climate change.

Part of my fascination with the Citizens' Assembly was with the
character of their leader, the retired judge who chaired the meetings.
A former judge of Ireland's High Court and then for four years its
Supreme Court, this eminent lawyer seemed surprisingly humble.[3] She
had already led her group through a gruelling inquiry into the Eighth
Amendment, overseeing an in-depth investigation into the medical
and ethical issues involved, during which citizens had taken detailed
evidence from church leaders, philosophers and medical experts.[4]
At the end of five weekends of deliberations, the group she was there
to support but not influence had recommended there should be a
referendum to decide the issue. When the government enacted the
Assembly's recommendation, the public referendum resulted in a clear
decision to repeal the Eighth Amendment. All this had been achieved
without the rancour and division my own country was going through
over Brexit. That seemed a remarkable achievement.

Judge Laffoy clearly had a fine legal mind but she also thought as
much about the people as the issues involved. In the bar the night
before, she had been solicitous of my welfare on my short stint serving
the Assembly and had expressed her gratitude for the contribution
I was about to make. And listening to her introduce this new topic
of climate change the following morning, I could see how she cared
about the welfare of the members of her Assembly and how this must

have helped steer them through the turbulent waters of the abortion debate. She seemed well capable of guiding them through this climate debate too.

After Judge Laffoy's brief preamble, the inquiry proper began with two presentations from Irish scientists on the causes and effects of climate change. The presentation following the coffee break was given by Joseph Curtin, a research fellow at a Dublin-based think tank on international affairs. He set out how international climate policy had been transformed since the time of the IPCC's 2013 report, the one I had helped to write.[5]

It had taken almost two decades, he explained, since the successful negotiations that produced the Kyoto Protocol before countries were able to agree on a treaty to succeed it. Following the debacle of the failed negotiations in Copenhagen in 2009, a new agreement was finally forged in Paris in 2015 thanks to a newly determined US administration under Barack Obama and a willingness from developing countries to consider limiting their own emissions. As far as a collective commitment to tackling climate change was concerned, the Paris Agreement was a game changer.

Under the Kyoto Protocol, only developed countries had agreed to reduce their emissions. The rationale was that since their historical emissions were mainly responsible for the bulk of the global warming problem, they should be responsible for fixing it. Since then, it had become increasingly clear that the central aim of the United Nations Framework Convention on Climate Change – to avoid dangerous climate change – would only be achieved if developing countries were *also* required to keep their emissions in check. In Paris, all countries had signed up to an agreement that committed to keeping temperatures to 'well below 2 degrees Celsius above pre-industrial levels' and to 'pursue efforts' to limit temperature increases to below 1.5 degrees.

To achieve this, signatories had agreed to 'peak global greenhouse gas emissions as soon as possible' and 'to undertake rapid reductions thereafter in accordance with best available science.'[6] Although the pledges made so far to reduce emissions fell short of achieving the goal of keeping warming to well below 2 degrees Celsius, the agreement also required nations to continually raise their ambition by committing to more stringent emission reductions in future. To encourage this, a process known as a 'global stocktake' was envisaged. Every five years, progress towards reducing and then eliminating net carbon emissions by each of the Paris signatories would be assessed. These periodic stocktakes, it was hoped, would encourage countries to ratchet up their efforts on becoming carbon-neutral as low carbon technologies became cheaper.

With the global context explained, Joseph Curtin now brought his presentation closer to home. Ireland, he told us, was party to a collective pledge made by the European Union to reduce their emissions from 1990 levels by 20 per cent by 2020, by at least 40 per cent by 2030 and by between 80 and 95 per cent by 2050. Unfortunately, according to the charts Curtin presented, Ireland was far from being on track to achieve these aims, which even then were still not sufficiently ambitious to meet the Paris goal of keeping global warming to well below 2 degrees.[7] Irish emissions from building, transport and agriculture were only 12 per cent below 1990 levels by 2017, and to make matters worse, overall emissions were now increasing. Collectively, the EU were doing rather better: by 2020 emissions would be over 30 per cent lower than 1990 levels and the UK's emissions would have dropped by 40 per cent by 2020 compared with 1990.[8] Far from being a leader on tackling climate change, Ireland was lagging badly behind. To meet its targets, the Irish economy would need to undergo a structural transformation which would have profound implications for the various economic sectors, jobs and society.

Questions from the Assembly members followed this sobering conclusion. They were all on the theme of why Ireland wasn't doing more to meet its international obligations on tackling climate change. Why wasn't there more investment in wave energy and rail infrastructure? Could we learn anything from New Zealand, a country with a similar population and dependence on agriculture? Could politicians use adaptation measures as an excuse to water down mitigation plans? Would there be any merit in a constitutional change to incorporate a stewardship obligation on politicians?

Wind energy was more cost-effective than wave, Curtin replied, and should be considered first. Ireland's dispersed population meant the economics of providing a viable rail network was very challenging. Even so, spending two thirds of its capital budget for transport on road was completely incompatible with a country that aspires to be carbon-neutral. As far as agriculture was concerned, the Common Agricultural Policy wasn't helpful. But being part of the EU put a lot more pressure on Ireland to reduce its rural emissions than a country like New Zealand. While any government could choose to prioritize adaptation over mitigation, or vice versa, making either choice would be a mistake as both needed to be taken seriously. Recent floods demonstrated the urgency both of becoming more resilient to such weather extremes and of taking action to lower their frequency in future by reducing emissions. A constitutional change to oblige the government to protect the environment would not be unprecedented in world affairs. Sweden, for example, had introduced such a measure.

Lunch was provided in the hotel's restaurant where I sat with a man whose dedication to the Assembly process was striking. He had arrived in the early hours, after a long drive across the country following the Friday night shift at the pub he owned in the south-west. Like most people here, he had already undertaken five intensive weekends on abortion and was now tackling climate change. As he talked about the task ahead, I was struck by how he approached his involvement. For him, far from being an imposition, he saw it as a privilege and a responsibility.

My presentation to the members was the first of the afternoon.[9] To set the scene, I showed a map of rainfall caused by Storm Desmond, an extra-tropical cyclone that in December 2015 transported a plume of moist air, known as an atmospheric river, to the south and west of Ireland and the west of the UK. The monthly totals showed over three times the average rainfall over large swathes of both countries. In its wake came devastating floods, closed roads and millions of euros worth of damages. What was the evidence that this, and other recent instances of extreme rain, heat and wind, had anything to do with climate change?

To start answering this question, I showed the members of the Assembly an illustration taken from the IPCC report I'd helped write. It demonstrated that the observed global mean warming of about 1 degree Celsius was entirely consistent with climate models that included increasing greenhouse gas concentrations and other human factors, but was completely inconsistent with climate models that included only natural factors. As we had summarized in the report, agreed in consensus by all governments in Stockholm, 'It is extremely likely that human influence has been the dominant cause of the observed warming since the mid-20th century.'

A gradual warming shift in the distribution of temperatures from day to day and month to month, I explained, implies a much more dramatic increase in the risk of extremely warm temperatures. What would once have been a very rare occurrence of extreme heat in the far tails of the distribution of a 'normal' range of temperatures, quickly becomes much more common, even under relatively modest increases in mean temperature. With global warming, the odds of extreme heat locally were shortening all the time. And the same thing was happening with rainfall extremes.

A warming atmosphere, I explained, contains more moisture as water is evaporated from the oceans. Rising at a rate of about 7 per cent per degree of warming, this increased moisture provides more energy to fuel storms and a dramatically increased potential for heavier rainfall. This does not necessarily result in the risks of extreme weather increasing in the same way everywhere. Varying ocean currents and atmospheric storm tracks can enhance such risks in some places more than others. These variations are hard to predict and may or may not be linked to the effects of climate change. But the thermodynamic changes from the warming atmosphere generally dominate. As a result, the risks of heatwaves and storms are increasing around the globe, and at a rapid rate due to increasing temperatures and rising atmospheric moisture.

When storms do form, they are more likely to be more extreme. And with sea levels rising at a rate of over 3 millimetres a year, there is an increased risk from storm surges and coastal inundation. Citing a piece of research by the hero of the early hours' debate in Stockholm on cumulative emissions, Reto Knutti, and his PhD student Erich Fischer, I showed a map of how the probabilities of hot extremes and heavy rainfall were changing across the globe.[10] The probability of extreme temperatures had already doubled in north-west Europe with 1 degree Celsius of global warming. If warming reached 2 degrees, such extremes would become five times more likely. The probability of rainfall extremes was also increasing appreciably. The greater risks of heatwaves and heavy rainfall were already becoming manifest.

I finished by making a point about avoiding the impacts of dangerous climate change. The pledges made in the Paris Agreement and the commitment to raising ambition through the global stocktake would limit the severity of climate change for huge numbers of people. Displaying an analysis made at the Met Office, I pointed out that many tens of millions would avoid flooding and many hundreds of millions would avoid water shortages each year if global warming were reduced from the 5 degrees Celsius of warming by the turn of the century projected under continued emissions to the 2 degrees of warming envisaged by the Paris Agreement. As far as climate action was concerned, there was still plenty to play for.

For the Q&A with the members, I shared a stage with Saji Varghese from the Irish Meteorological Service, Met Éireann, who had followed me with a talk describing how climate was changing in Ireland.[11] Temperatures had warmed by just under 1 degree Celsius in the country, he explained to the members of the Assembly, and the number of days with heavy rainfall of more than 10 millimetres had increased.

Projections from climate models showed continued greenhouse gas emissions would bring the Irish an increasing frequency of extreme temperatures and more and more heavy rainfall events. Ireland was sharing the same fate as I'd mapped out for the globe more generally – a warming world increasingly prone to damaging heatwaves, floods and storms.

As before, the Assembly members asked questions of the panel. The ones they addressed to me seemed particularly pertinent. 'As far as climate change was concerned, what was the worst that could happen?', they asked. 'What was the best?' 'And if I was in charge, what was the one thing I would do to change things for the better?'

These were not the sorts of queries I was used to dealing with at academic conferences or in government briefings, nor had I any answers pre-prepared. With little time to reflect, I fell back on my gut instincts, informed by my many years of research and my recent experiences at home. The responses to the best and worst could be related to the potential trajectories of climate models. The response to the world-leader-pretend question could reflect my own personal struggles with trying to go green.

A global warming of 5 or 6 degrees Celsius, the levels anticipated within a century if greenhouse gas emissions continue to increase, I told the Assembly, would tip the world into a state of collective desperation. Crops would be decimated by drought, sea level rise would inundate tens of millions, and water shortages would cripple whole countries. And this would only be the beginning of the hell to come. At such levels of warming, the Greenland and West Antarctic Ice Sheets could be undermined beyond repair, unleashing many metres of sea level on future generations, and the earth's permafrost could melt catastrophically which would release vast quantities of methane, thereby turbocharging global warming even further. Given that the last ice age was only 5 or 6 degrees colder than now, it should be no surprise that a 5- or 6-degree warming would unleash a similarly vicious reality, albeit a hothouse earth rather than a freezer one.[12] It would be a planet on which many societies, with all their complex interdependencies, would struggle to survive.

The best we could hope for was the 1.5 and 2 degree Celsius worlds envisaged in the Paris Agreement. Limit global warming to such relatively modest levels and we could anticipate a changing climate that with suitable adaptation could avoid catastrophe for many. The 2003 heatwave had killed 70,000 people across Europe but lessons had since been learned. Subsequent heatwaves had been coped with much better, at least in richer countries, thanks to greater warnings and provision of care for the elderly and vulnerable. Inhabitants of low-lying island states might still have to evacuate and the Arctic people would still face a landscape changed beyond recognition. But the best that could be hoped for was a huge amount better than the worst. With renewable energy powering our economies rather than fossil fuels, children could once again be breathing clean air, which would be a huge co-benefit of climate change mitigation.

And then we came to what I would do if it were me in charge of fixing climate change. My answer was reported by George Lee, the environment and science correspondent for Ireland's national broadcaster, during that evening's television news bulletin. Reporting from the meeting he said, 'The one hundred members of the Citizens' Assembly listened with increasing concern as expert after expert outlined to them today the clear evidence of how climate change is impacting, the consequences if we do nothing about it and the challenge Ireland faces just to live up to its responsibilities and play our part, never mind to becoming leaders in tackling climate change.'

'The mood of the Assembly might well have been revealed,' he summed up, 'when the one question was put to the panellists about what one thing would they do to tackle climate change, and Professor Stott of the UK Met Office said he would make it far easier for citizens to access the grants and support to do the right thing such as installing solar panels and heating systems. For that he drew the only spontaneous round of applause of the day.'[13]

I had clearly struck a chord. For over a year, we had been trying to initiate a renovation project of our house, one that involved installing solar panels and a new more energy-efficient heating system. But the complexities of the green aspects had prevented us finding a builder prepared to take it on. The capacity to undertake such work in our region had been slashed after the UK government withdrew subsidies aimed at increasing the use of renewables.[14] With the need to tackle climate change becoming ever more urgent, it seemed absurd that governments like mine were making it harder, not easier, for people to take environmentally beneficial decisions.

In Ireland it seemed, things were even worse. While the UK did have a feed-in-tariff for domestic generation by solar panels or wind turbines at the time, although this was stopped in 2019, Ireland had never had such a scheme.[15] Confronted by the seriousness of climate change, the Assembly members were frustrated at the difficulties put in the way of making a positive contribution. Reducing your personal carbon footprint could be infuriatingly difficult and expensive, whether by home improvements, travel to work or choice of diet. My story about my own difficulties in doing the right thing had unleashed an upwelling of collective frustration from the Assembly members.

That evening, we all had dinner together in the hotel restaurant. The food was excellent, the company friendly and the mood relaxed. Now that I'd given them my presentation and answered their questions on the panel, the members welcomed me as part of the gang, no longer a stranger but a fellow comrade to the cause of evidence-based policy advice. After dessert, in what had apparently become a regular Saturday night Assembly tradition, one of their number sang us some of his folk songs, his rich baritone voice filling the venue.

The following morning, I returned home while the Assembly continued its investigations into how Ireland could become a leader in tackling climate change. The members met again for a second weekend in early November, on the Sunday of which they voted on a number of propositions, the nature of which had been decided upon by the members themselves. The outcome was revealing.[16] The citizens were drawn to represent the demography of their country in terms of age, gender, social class and regional spread. Yet, in voting, they expressed a very strong majority view on every single motion.

Ninety-seven per cent of the members recommended that a new body should be set up to ensure climate change was at the centre of policymaking. One hundred per cent of the members recommended that the State should take a leadership role in addressing climate change through mitigation measures, including, for example, retrofitting public buildings, having low-carbon vehicles, renewable generation on public buildings, and through adaptation measures such as increasing the resilience of public land and infrastructure. Eighty per cent of the members stated that they would be happy to pay higher taxes on carbon-intensive activities.

Further votes drilled down into the details: 96 per cent of the members recommended that the State should undertake a comprehensive assessment of the vulnerability of critical infrastructure, 99 per cent wanted the State to enable the selling back into the grid of renewable energy micro-generation by private citizens, 100 per cent supported the greatest possible community ownership of future renewable energy projects and 97 per cent agreed that there should be an end to subsidies for peat extraction. In addition, 93 per cent voted for an increase in bus and cycle lanes, 96 per cent agreed that the State should support the transition to electric vehicles and 92 per cent supported an expansion of public transport. Furthermore, 89 per cent were in favour of a tax on greenhouse gas emissions from agriculture, 93 per cent wanted mandatory reporting of food waste in the food distribution and supply chain, and 99 per cent voted for measures to encourage tree planting and organic farming.

Later, looking over the Assembly's final report, I could see why the members had reacted the way they had, given the evidence put before them. Much of what they learnt they would have found surprising and thought-provoking. I wasn't aware, for example, that international experience had shown that allowing people to become energy producers themselves was a major catalyst of engaging people on climate change, or that food waste accounts for 8 per cent of global greenhouse gas emissions, or that Scotland had shown it is possible to make a rapid energy transition, having increased their renewable energy generation from 10 per cent of the total to 60 per cent in the last fifteen years. These were facts that the great majority of people would normally never come across. When confronted with information like this, a cohort of randomly chosen people from all walks of life had come to a clear view, not just on the rapid change of course their country needed to become a leader in tackling climate change, but also on the realistic possibility of doing so. They would support this change of course, even at the cost of higher taxes on polluting ways of life. That in itself was a surprising fact.

My own country, the UK, was in some ways far ahead of Ireland. The Climate Change Act of 2008 committed by law the government to reducing greenhouse gas emissions in line with recommendations of the independent group of experts on the Climate Change Committee.[17] As part of the legislation, these stringent legally based targets would also be subject to revision later if there was new evidence to take account of. This was just the sort of framework Ireland was crying out for and which was also lacking in most other countries of the world. In some respects, at least, my own country was a world leader.

In other respects, it wasn't. Returning to my own shores in September 2017 from meeting the engaged, well-informed and committed members of the Citizens' Assembly of Ireland, I couldn't help thinking such a process would help matters considerably back home. Over here, the government's aim to reach net zero emissions still needed to be put into practice when many people seemed disillusioned or apathetic about climate change. Public engagement had never recovered from the battering it had received following the Copenhagen debacle and the Climategate affair of 2009.[18] Almost a decade on, the relatively sparse coverage in the media of the issue seemed vastly out of proportion to its importance. Television and radio documentaries largely avoided what was seen as a turn-off issue. When the BBC did deal with it, they were still prone to making the old familiar errors involved in false balance.

On 11 August 2017, I had appeared on Radio 4's *Today*, the BBC's flagship current affairs programme.[19] The previous day's edition had included an interview with Nigel Lawson in which the former Chancellor of the Exchequer and founder of the climate change denial Global Warming Policy Foundation had claimed that global temperatures were declining and that 'all the experts say there hasn't been' an increase in extreme weather events.[20] When the inevitable outcry broke out, the BBC asked me to go on the programme the following morning to respond. Given that both of Lawson's statements were false, I was happy to rebut them on behalf of the climate science community. Unfortunately, my rebuttal came during another edition and at an earlier time. Many people who caught the original fallacious claims would have missed it. The episode felt like yet another setback in the greater cause of preventing dangerous climate change.

Climate scientists in other countries also faced difficulties in getting their messages across about the mounting threats from global warming. After announcing during his first year in office that he intended to withdraw the US from the Paris Agreement, the following year President Donald Trump denounced the country's National Climate Assessment, the work of thirteen federal agencies.[21] 'I don't believe it,' he said of a report that concluded that 'Earth's climate is now changing faster than at any point in the history of modern civilization, primarily as a result of human activities.'[22] His disregard for scientific evidence went hand in hand with his efforts to roll back many of the environmental regulations introduced during the Obama years.

In Australia, Malcolm Turnbull's administration blew hot and cold on climate change, adopting the Paris Agreement but blaming electricity blackouts on over-ambitious renewable energy targets.[23] In response to the government's decisions to cut grants for environmental research, in 2016 Australia's leading research body, the Commonwealth

Scientific and Industrial Research Organisation (CSIRO), announced a severe cutback of its climate change work with the loss of up to 350 staff.[24] Invited to speak at the annual meeting of the Australian Meteorological and Oceanographic Society in Melbourne, I had taken part in a walkout by scientists at the conference to protest about the cuts at CSIRO and spoke to Australian media about the international excellence of work being carried out there. It was absurd that world-leading and world-saving research was being sacrificed at such a critical juncture in global efforts to tackle the climate crisis.

*

It was hard to see how progress would be made, without countries following the lead of Ireland and having processes like Citizens' Assemblies where ordinary people could engage with the scientific evidence and spur their governments to take it seriously. The UK might have long-term commitments to reducing emissions but these were set for timescales much longer than the short-term cycles of domestic politics. With Brexit absorbing much of the government's bandwidth, where was the appetite to implement the recommendations of the Climate Change Committee when there was little sign of a growing public clamour for action? What would it take to turn the situation round?

Just like in Ireland, the British were seeing their fair share of weather extremes. In January 2014, the then prime minister David Cameron had visited the flooded Somerset Levels in his wellington boots during one of the wettest winters on record. Later, during Prime Minister's Questions, he declared that he suspected that the extreme rainfall which caused disruption and the loss of several lives was linked to climate change, even though his Environment Secretary, Owen Paterson, was a climate-sceptic-leaning politician averse to making any such connections. Two days later, Myles Allen and I gave a briefing to a panel of journalists on the recent weather and climate change. There was only one question they wanted to ask us. Was the PM correct in his assertion linking the flooding to global warming?

Myles and I agreed that he was correct in suspecting such a link.[25] We had more research to do to understand exactly how increased moisture had combined with the varying jet stream to fuel this particular heavy rainfall event. But we were confident that the weather was changing, bringing with it a greater risk of heatwaves, droughts and floods.

The following years would bear out our prediction. A succession of heatwaves in Europe culminated in the extreme summer of 2018 with record-breaking temperatures both in the UK and in many other places around the northern hemisphere.[26] Extreme heat in Japan and Korea caused hundreds of deaths, wild fires raged out of control in California destroying thousands of homes, severe flooding led to the displacement of tens of thousands of people in Myanmar, and drought in Scandinavia led to the substantial loss of crops.[27] If the people weren't listening to the scientists, nature was sending out an increasingly alarming message.

At the end of that summer, the hottest in her country since instrumental records began, a fifteen-year-old girl stayed away from school and sat outside the Swedish parliament with a home-made sign saying 'Skolstrejk för klimatet' (meaning 'school strike for climate' in English).[28] To begin with, on that first day, she was on her own and nobody, apart from herself, her family and her teachers, cared. But later that day, her stance had been picked up on social media and on her second day of striking, the following Friday, she was no longer alone. Within just six months there were children striking all around the globe calling on their governments to take heed of the science and start addressing the climate crisis. Greta Thunberg's school strike had started something big.

The day I followed the school strike marchers in Exeter city centre, on Friday 15 March 2019, there were an estimated 1.4 million young people marching in 123 countries.[29] This new level of engagement after so many years of widespread apathy was truly heartening. Judging by the placards held up by the large group of youngsters walking along my local high street, a new generation didn't want the apathy of their elders to carry on any longer. 'The Climate Is Changing', said one, 'Why Aren't We?'

By then, the scientific community had thrown another log on the fire of rising public concern. With the global climate agreement in Paris in 2015 came a request to the IPCC to produce a special report on the central aim of that agreement – to keep global warming to well below 2 degrees Celsius above pre-industrial levels and to pursue efforts to keep warming to below 1.5 degrees. The policymakers wanted the IPCC to compare the impacts of 1.5 degrees warming with 2 degrees and to set out the emission cuts needed to limit warming to such levels.

The IPCC responded with remarkable speed in producing a report that involved ninety-one scientists from forty countries assessing over 6,000 peer-reviewed scientific papers.[30] Released on 8 October 2018, it provided stark testimony about the damage to be faced with just an additional *half a degree* of warming. Sea level, extreme heatwaves, heavy rainfall and drought would all increase. The risks for ecosystems, food and water security, health, development and economic growth would also mount. But such risks would be substantially lower at 1.5 degrees of warming than at 2 degrees. As the IPCC put it in a video to accompany the report, 'every bit of warming matters'.[31]

The IPCC report also stressed the crucial nature of urgency in tackling climate change. Warming had already reached 1 degree relative to pre-industrial levels. Limiting warming to 1.5 degrees meant that emissions of carbon dioxide would need to halve by 2030. They would then need to be eliminated entirely by 2050. Deep cuts would also be needed in emissions of methane and other greenhouse gases.

Behind these headline conclusions lay some simple yet devastating mathematics, first pointed out by Myles Allen.[32] The basic arithmetic is a straightforward consequence of his landmark finding that global warming is directly related to cumulative emissions of carbon dioxide. This relationship means that to stop warming, CO_2 emissions must also stop. With global warming already at 1 degree Celsius, stabilizing warming at 2 degrees implies CO_2 emissions have to reduce by 10 per cent for every tenth of a degree of warming. Such a rate of reduction would allow emissions to fall to zero by the time warming has reached a further degree. To stabilize at 1.5 degrees, emissions would need to reduce by 20 per cent for every tenth of a degree of warming. The maths showed that the task of limiting warming was theoretically very simple. But it also showed, practically speaking, that to achieve it was going to be dauntingly hard.

Such a task *was* possible, the IPCC concluded, if there were a steep decline in coal use, greatly increased energy efficiencies, the development of new technologies and behavioural changes.[33] Doing so posed massive challenges. They could be overcome but only if there was the political will to accelerate the transition to a greener and more sustainable economy.

In laying out the science so clearly, the IPCC had helped focus minds, like Greta Thunberg and the school strike movement had, on the hopes and fears that global warming presented. Keeping further warming to less than another degree was still possible. But politicians needed to act fast and societies needed to embrace the challenge.

With little time to waste, at least the meteorology was not hanging about. With ever more heatwaves, floods and storms around the world, scientists needed to act fast themselves and work out what all this changing weather meant, both for now and for the future.

12

Change is coming

All scientific research is a long-term endeavour. But increasingly in our scientific community we were butting up against the escalating rapidity of the passage of events around us. Events in nature and events on the street; they were starting to happen at a pace that meant the status quo was rapidly becoming unsustainable. Scientists couldn't wait several years for new findings to emerge before making their assessments of what was happening around them. Nor could politicians risk delaying climate policies to tackle the mounting threats of climatic disruption for yet another electoral cycle. Change was coming, whether we liked it or not, and fast.

That phrase, from Greta Thunberg – 'change is coming, whether you like it or not' – had really struck a chord when I heard her speech at the United Nations Climate Change Conference in Katowice, Poland, in December 2018. In the past, I had sometimes thought of climate change as slow-moving and largely in the gift of powerful politicians to do something about it. In her address to the representatives of the powerful, she turned this argument on its head. In the face of a rapidly diminishing window of opportunity to forestall climate catastrophe, she told the world's leaders that the familiar state of affairs was coming to an end.

'We have not come here to beg world leaders to care,' she told them. 'You have ignored us in the past and you will ignore us again. You have run out of excuses and we are running out of time. We have come here to let you know that change is coming whether you like it or not. The real power belongs to the people.'[1]

I had travelled to the coal-mining town of Katowice to present some of our latest scientific results to delegates at the conference. The meeting in Poland, the twenty-fourth annual Conference of the Parties to the United Nations Framework Convention on Climate Change (COP24), was the sixth that I had attended. My first was in Kyoto in 1997 where, for the first time, world leaders had agreed a Protocol to start reducing greenhouse gas emissions. Cath Senior and I, the two junior representatives sent east by our bosses at the Met Office Hadley Centre, presented our results on planetary warming from a table in the main lobby of the compact conference centre. Already by that year, the hottest so far in the instrumental record of global temperatures, climate research had identified the distinctive fingerprint of human activity in atmospheric warming and had shown that further warming would lead to increased hunger for millions, water shortage for many and devastating floods for inhabitants of the world's coastlines. Politicians knew enough to act, even then.

At each of the COPs I've attended since, the atmosphere has felt different. In Kyoto, there was a fevered air of excitement as negotiators raced towards their historic agreement. But without much progress towards the targets for reducing emissions set in Kyoto, the atmosphere at the meeting I attended in Milan in 2003 felt listless and flat. Two years after the catastrophic failure of the Copenhagen meeting in 2009, I travelled to Durban where the EU chief negotiator, Connie Hedegaard, worked miracles at re-energizing a moribund process.[2] The mood in Warsaw in 2013 felt more hopeful and the following year in Lima I thought I detected a renewed purposefulness among the suited ladies and gentlemen hurrying from one tented structure to another on the converted military parade ground where the meeting took place. With countries working towards the Paris showdown the following year, it felt like the tide was finally turning.

So it seemed when the Paris Agreement was approved on 12 December 2015 and greeted around the world as a major stepping stone towards preventing dangerous human interference with the climate system. But in Katowice in 2018, the mood in the vast conference complex – a huge congress centre and indoor sports arena connected by a network of corridors, tented rooms and double-decker exhibition halls – seemed to have switched again. It now felt fractious and edgy. Tens of thousands of people – negotiators and journalists, lobbyists and scientists – had gathered in this cold and drab city for a meeting intended to agree the 'rule book' by which the Paris aims would be met.[3] With the latest data showing global greenhouse gas emissions reaching record highs, and demonstrations by groups like Extinction Rebellion showing that people were starting to lose patience, the Paris aspirations to keep global warming to well below 2 degrees Celsius had to be turned into progress on the ground and fast. Doing so, and without further delay, was starting to feel like a worryingly pressing imperative.

I spent most of my time in the huge two-storey tent set aside for the side-events hall, which was where I presented my scientific results. This aspect of the conference had grown massively from those early days in Kyoto in 1997 when attending a side event on the science meant talking with Cath and me at our table in the foyer. By 2018, finding solutions to the climate crisis had become a huge international concern. Countries had their own pavilions with daily programmes of presentations on science, technology and societal change. The side-events hall was loud, bright and crowded with people from morning until night.

At the British pavilion, which came equipped with rows of benches in front of a large television screen, I presented a new analysis that Nikos Christidis in my research team and I had made of that year's record-breaking heat in the UK.[4] It was a study that followed in the footsteps of the pioneering assessment I had made of the 2003 European heatwave, which was published in *Nature* in 2004. Since then, we had developed the capability to carry out these types of studies much more quickly. Rather like weather forecasting, we could now apply well-established methods to guide people about heat and cold, floods and droughts, without having to wait for an academic paper to be published in a specialized journal. Rather than predicting the weather over the next few days, we were advising people about how weather that had recently happened had been altered by human-induced climate change. We called this operational attribution.

The basic principle behind our analysis of the 2018 heatwave was the same as behind my analysis of the 2003 heatwave. We estimated the statistics of temperature extremes in the current climate from a set of climate model simulations with elevated atmospheric concentrations of carbon dioxide, other greenhouse gases and human-induced pollutants, and compared them with the statistics of temperature extremes from a set of climate model simulations with pre-industrial levels of greenhouse gas concentrations and no other human-induced pollutants. By comparing the statistics in these two different situations, the unaltered natural world and the altered human-affected one, we could make a calculation of the altered probability of the extreme temperatures seen that summer.

The results showed that human-induced emissions had increased the chances of the summer heat of 2018 by about thirty times. The odds of such an extremely warm summer would once have been about 0.4 per cent each year, but had now risen to about 12 per cent. That was quite a shift.

The findings attracted plenty of interest from journalists, including from the BBC's correspondent Matt McGrath who interviewed me for the television and radio news. That spring, the BBC had been found in breach of broadcasting rules by Ofcom for its controversial Radio 4 *Today* programme interview with Nigel Lawson in which he claimed that global warming had stopped and there was no increase in extreme weather events.[5] In September, the BBC had released formal guidance to its journalists on how to report climate change which included acknowledging that human-induced climate change exists and the need to avoid false balance. I was pleased to know they had done this, even though excellent journalists like Matt McGrath and David Shukman, with their long and distinguished records of reporting on climate change, had no need of such advice. The BBC had recognized that some of their editors and non-specialist reporters did. It was another encouraging sign that the media landscape was shifting in favour of more balanced reporting that recognized the urgency of the climate crisis.

The following spring, the BBC broadcast a one-hour climate change documentary on BBC1 fronted by Sir David Attenborough. This was a remarkable occurrence – the first time the national broadcaster had aired a primetime documentary on the subject since the pre-Climategate days of 2007 and a programme also presented by Attenborough called *Climate Change: Britain Under Threat*.[6] This new programme for 2019 took a more global perspective and a back-to-basics approach. In a one-hour documentary it set out the science, impacts and potential solutions of climate change. It was called *Climate Change – The Facts*.

I had been asked to take part and was asked question after question about the nature of climate change, its causes and the likely threats if emissions continued unabated. It was a fascinating process to be involved in, with the director/producer Serena Davies and I developing a sense of collaboration as we searched for the clearest, most concise way of phrasing a particular point I wanted to make. But after a long and tiring day, I was glad when Serena finally declared she had what she needed and she and her team could set off back to London.

While David Attenborough would introduce the film, link the sections and bring in a mass audience, Serena's aim for the programme was to have the story largely told by its protagonists, people caught up in terrifying fires and ecological disasters, scientists who had been warning of the threat for years, and engineers developing renewable forms of energy generation. She had also recorded an interview with Greta Thunberg as part of the final section focusing on what individuals can do in terms of lifestyle changes and political lobbying.

Telling the story this way, rather than by a voiceover intercut with sound bites, took more time to put together. I knew my science well enough, why the climate was changing and what was happening to heatwaves, floods and droughts. But engaging a broadcast audience of millions was a different challenge from the more familiar ones of talking to my peers or addressing a public audience of interested weather enthusiasts. By asking me the right questions and listening to the thread of argument, then helping me craft a clearer answer that left out inconsequential digressions but brought out the implications, Serena was able to gather the material together for editing later. This was an enterprise that depended for success on a trusting relationship between interviewer and interviewee.

It was an enterprise that seemed to pay off. The critics were positive, the audience figures large, and it made an important contribution to a rising interest in climate change. In the days after its transmission, almost everyone I spoke to mentioned they had seen the programme. A neighbour said that it had clarified the main issues for her about the climate crisis. A fellow tennis player who has two young children, thinking about the sequence where a father and son narrowly escape a raging inferno, said climate change made him feel scared. A friend said that watching the documentary had inspired her to go online immediately and switch both her own and her elderly mum's electricity supplier to renewable energy.

By May 2019, after the hottest Easter Monday on record for each of England, Scotland, Wales and Northern Ireland, analysis was showing a rapid change in attitudes.[7] Thanks to the school strike movement, the protests by Extinction Rebellion, the visit of Greta Thunberg to the UK parliament in the last week of April and the BBC1 documentary, media coverage of climate change had undergone a massive surge, surpassing its peak around the time of Climategate and the Copenhagen climate talks debacle.[8] At long last, the issue was back at the top of the news agenda, but this time framed in terms of urgency and action rather than delay and indecision.

That same month, the UK's Climate Change Committee released its most significant report yet.[9] Taking account of the latest scientific evidence provided by the IPCC's Special Report on Global Warming of 1.5°C, the committee recommended a new emissions target for the UK. This was that emissions of greenhouse gases should have reached net zero by 2050. This target, which was subsequently accepted by government, would deliver on the commitment the country made by signing the Paris Agreement, would be achievable with known technologies and would improve people's lives. However, it would only be possible, the committee advised, if further policies to reduce emissions were introduced without delay.

With alarm bells ringing in the human sphere, the sirens in the geophysical world continued to wail ominously. For the first time in the satellite era, sea ice was at record low levels at both poles simultaneously. In the Arctic in early June, as it melted towards its September minimum, ice coverage was already more than 1 million square kilometres below its 1981–2010 average.[10] In the Antarctic, as ice recovered towards its seasonal maximum, ice coverage was the lowest recorded in June since the start of the satellite record in 1979. The tropics too were feeling the heat. In the lead-up to the monsoon, India was scorched by an intense and prolonged spike in temperatures. Delhi reached 48 degrees Celsius, its hottest June day on record.[11] And in Europe, after the continent's record-breaking temperatures the year before, we waited to see how this summer was going to pan out.

On 23 June 2019, I travelled to the home of Météo-France in Toulouse for the Fourteenth International Meeting on Statistical Climatology.[12] Taking place every three years at different locations around the world, this is the foremost international conference on the applications of statistical methods in climate science. At the start of the meeting, we were shown the latest weather forecast which was for it to be extremely hot.[13] Temperatures were predicted to soar, climbing sharply to well over 35 degrees Celsius later in the week with a chance that they could peak at over 40 degrees on Thursday and Friday. You didn't need to be a meteorologist to know that temperatures like these would be difficult to cope with, especially for the vulnerable and elderly.

I had seen in the local paper that local authorities were planning for unprecedented heat and already journalists were ringing our press office asking us to comment about a possible climate connection. This heightened interest in weather and climate change, even in advance of the extreme weather arriving, was yet another sign of a growing public awareness of the harmful effects of humanity's greenhouse gas emissions. It was a coincidence that many of the experts in attribution had gathered together at the heart of the predicted heatwave. But conscious of our collective responsibility to respond to media interest, a group of us decided to calculate the extent to which it had been affected by human causes. In another application of operational attribution, we were going to do so fast, within just a few days.

The analysis was led by Fredi Otto from Oxford University and Geert Jan van Oldenborgh from the Royal Netherlands Meteorological Institute (KNMI), who had pioneered a fast-track approach to the attribution of weather events in a project called World Weather Attribution.[14] They had already shown the feasibility of making attribution assessments in just a few days as long as the meteorological data were readily available and the scientists were able to drop everything and make the calculations. Located as we were on the campus of the national weather service in France, on this occasion we had everything we needed to carry out the analysis.

Over the next few days, as the group started analysing the data while also attending sessions of the conference, the weather built towards its climax with a strong hot wind blowing across the parched grass outside. Leaving the air-conditioned interior of the conference centre felt like entering a fan oven. This was no cooling breeze, but wind straight from the Sahara. It set a suitably ominous backdrop for our attribution task team which included researchers from seven different institutes – the KNMI, the Institut Pierre-Simon Laplace in Paris, the University of Oxford, Météo-France, the Swiss institute ETH, the Red Cross Red Crescent Climate Centre, and the UK Met Office.

Collectively, we investigated the evolving meteorological structure of the heatwave. The weather charts showed a low-pressure system forming off the coast of the Iberian Peninsula and funnelling air northwards from North Africa directly towards southern France. We decided to look at three-day average temperatures, the timescale over which the heatwave was expected to be at its most intense. If records were broken later in the week, we could put the observations alongside the simulations of our climate model and work out how much more likely these new records had become thanks to climate change. Then, when we had double-checked our calculations and written up what we had done in a short report, we could present our results to the media early the following week.

The heatwave peaked in Toulouse on Thursday. The temperature indicated by the LED display at the front of the coach was 41 degrees Celsius as we drove off the Météo-France campus at 6.30 p.m. When I returned to my lodgings at just after 10 p.m. after the conference social event sailing sedately down the Canal du Midi, the thermometer outside my hotel showed 34 degrees. To be sure about the day's highs, we would have to wait until the official figures came through from Météo-France. But that night, scanning through the available weather station data online, it looked like Toulouse had just broken its June record with a maximum temperature of 40.2 degrees Celsius. Now the question became, would the all-time temperature record for France be broken the following day?

The next day, France did break its all-time temperature record, not just for June but for the whole summer and by a massive margin. The weather station at Gallargues-le-Montueux, a village near the city of Nîmes in the south-east of the country, recorded 45.9 degrees

Celsius, a whopping 1.8 degrees warmer than the previous French record of 44.1 degrees.[15] This was not just unprecedented. This was a temperature that would have been inconceivable to the people of the region just a generation before. Forty-five degrees could never happen in my country they would have said. But now it had.

That last week of June, temperatures across much of Western Europe were between 6 and 10 degrees warmer than the climatological June average.[16] This was so extreme that we couldn't be sure our methods would be capable of evaluating the statistics correctly and we had much frantic checking to do of the results calculated from different climate models. Large uncertainties remained. Nevertheless, all the results pointed to a substantial increase in the probability of such extreme temperatures.

The following Tuesday we released our findings to the media.[17] We concluded that human activities had made the heatwave at least five times more likely, although it could be many more times than that. In future, such heatwaves would become commonplace if greenhouse gas emissions continued on their current trajectory. These figures were shocking. The abnormal was becoming normal at a frightening pace.

There was plenty of interest in these new results from the media. Never before had a scientific team like this produced an attribution analysis so quickly. Twenty-three years on from the IPCC's assessment of a 'discernible human influence on global climate', this obscure, life-changing and bitterly controversial scientific field of ours had just reached another major milestone. Only ninety-six hours after temperatures peaked, an attribution assessment had been made of the human influence on France's most extreme heatwave on record.

Having so many experts gathered together at the time and place of France's record-breaking heatwave made this a perfect storm for climate science. It was a storm of heat that gave us a chance to apply our scientific ideas as the extraordinary weather built around us. It was a storm that provided us with a unique platform to raise awareness about the impacts of continued emissions. And mercifully, it was a storm without the terrible repercussions of the 2003 heatwave which had killed 70,000 people. The weather warnings were better now and the authorities better prepared. Luckily too, the heat abated much more quickly than in summer 2003, the super-heated weather sinking back to its subtropical origins in Northern Africa. Next time, if the heat stayed longer, Europe might not be so lucky. Vulnerable people who could manage a night or two of extreme heat might not survive if the super-heated weather lasted much longer.

Many other parts of the world were having it much harder.[18] Devastating floods during India's wettest monsoon for 25 years killed at least 1,750 people. Hurricane Dorian, the second strongest Atlantic hurricane on record, ripped into the Bahamas and then North Carolina causing billions of dollars of damage.[19] And towards the end of Australia's hottest and driest year on record, the country suffered one of its most apocalyptic starts to the fire season ever with at least 21 people killed and 3,500 properties destroyed.

If the devastating consequences of climate change could not have been made any clearer by events on the ground, powerful politicians were still trying to forestall international efforts to tackle the root cause of all this mayhem. On 4 November 2019, Trump, making good on his promise to withdraw from the Paris Agreement as soon as the terms of the agreement allowed, formally gave notice of his country's withdrawal from a treaty he regarded as unfair and which he thought would decapitate his coal industry.[20] That year's COP meeting had been scheduled to take place in December in Brazil. But the new president, Jair Bolsonaro, rejected the scientific consensus on climate change, slashed funding for environmental research and favoured agribusiness over protection of the Amazon rainforest.[21] Unsurprisingly he pulled out of hosting COP25, which was then relocated to Chile before having to be moved again, just a month before the planned start, due to social unrest in the new host country.

Relocated a second time to Madrid, the results were hugely disappointing.[22] The success of the Paris Agreement depended on countries pledging to ramp up their reductions in emissions at a regular series of five-year iterations of the so-called ratchet mechanism by which ambition would be strengthened over time. With the first iteration due in 2020, the Madrid meeting was supposed to make progress towards this goal by countries announcing their intentions to make more ambitious pledges at the next annual COP meeting in Glasgow. But it didn't happen. The grouping of countries at existential risk from sea level rise – the Association of Small Island States – blamed Brazil, India and China for blocking progress. As before in Copenhagen in 2009, nations were not united in tackling the climate crisis but instead seemed bitterly divided.

During the meeting, a huge protest march including Greta Thunberg illustrated the yawning gap between the need for urgency demanded by the science and the snail-like progress being made by the politicians.[23] In her final speech to conference delegates, Thunberg captured the mood of many when she said that the meeting seemed 'to have turned into some kind of opportunity for countries to negotiate loopholes'.[24]

The outbreak of a global pandemic forced another delay to progress on climate action when the Glasgow meeting, originally scheduled for November 2020, had to be postponed until November 2021. No one was unaffected by Covid-19. I was relatively lucky in not falling ill and in being able to work from home – meeting colleagues, attending meetings and supervising my PhD student at the University of Exeter by video calls from my laptop. Sheltering in our house in Exeter under successive lockdowns, Pierrette and I followed the news of the world outside with grim fascination, as the death toll mounted and scientists raced to develop and then test vaccines that could one day bring an end to this planetary crisis. Pretty much everything seemed to have changed. But one thing that kept me busy certainly hadn't. That was the hectic pace of climate change.

2020 was shaping up to be another extremely hot year globally, following on from the previous six years as all being hotter than any year before 2014.[25] Carbon dioxide concentrations reached new heights in a record of bubbles of air trapped in Antarctic ice that stretched back 800,000 years.[26] The heat trapped in the oceans was greater than had ever before been measured.[27] And all of that extra heat was having more and more severe consequences for people living on land.

On 20 June 2020, the Russian town of Verkhoyansk recorded the highest ever temperature above the Arctic Circle, a sweltering 38 degrees Celsius.[28] It was the culmination of an extended period of Siberian heat that stretched back to the beginning of the year, more than 5 degrees warmer than average from January to June. Many of us who had analysed the French heatwave in Toulouse the year before decided to form a new team to make a rapid analysis of this striking event. We had not yet considered the Arctic regions despite their importance for people living at lower latitudes. Permafrost, snow and ice were all melting at a dramatic rate and these unprecedented changes could potentially be altering the jet stream and affecting European weather patterns.

The analysis was led by Andrew Ciavarella, one of the scientists in my research team at the Met Office and involved myself and researchers from six other institutes, including the Shirshov Institute of Oceanology in Russia. Applying the same methods we had applied to the French heatwave of June 2019 – comparing the current climatic conditions with those without human-induced emissions of greenhouse gases and other pollutants – we calculated the changed odds of such extreme temperatures in this most northerly region of earth. The results were astounding.[29]

Without the warming from greenhouse gas emissions, we found that the prolonged Siberian heatwave would only be expected to happen less than once in every 80,000 years. It would have been almost impossible. Now, thanks to human-induced climate change, it had been made over 600 times more likely. It was the strongest climate link to any extreme weather event that had ever been made. It was an extraordinary finding and one that I was keen should be widely known.

On the day of the press release, 15 July 2020, I went into the office for the first time in four months to record an interview with Justin Rowlatt, chief environment correspondent at the BBC. The clip was included in the BBC's lead story on *News at Ten* as part of its *Our Planet Matters* series.[30] The clear and unequivocal nature of the presentation felt a world away from some previous reporting that, as Steve Jones had pointed out in his review of the BBC's science output in 2011, had fallen prey to false balance. Now the viewer could be in no doubt about what science showed, that human influence on the climate system was having devastating consequences. As Huw Edwards said in his introduction to the news that night, our stark findings raised profound questions about the future of the earth's climate.

That future depended to a large extent on what happened in the US where there were some disturbing developments. In the run-up to the November 2020 election, the Trump administration removed the chief scientist at the National Oceanic and Atmospheric Administration (NOAA), the country's main scientific agency concerned with climate change, and replaced him with a former researcher for the libertarian Cato Institute who had criticized climate scientists for making overly dire predictions.[31] In another move designed to influence the next National Climate Assessment after the president had disavowed the previous assessment in 2018, a new senior post at NOAA was filled by a geographer who questioned the human-caused nature of global warming. These manoeuvres felt like chilling portents of what could happen if Trump was given four more years.

What further purges of climate scientists could there be, I wondered; what more attempts to criminalize our actions like I had seen in 2010 from Republican Senator James Inhofe. And what damage could there be, I wondered, to nations' efforts to raise their ambitions in reducing emissions towards the target of net zero by 2050, the target the IPCC had shown was needed to meet the Paris goal of keeping the effects of global warming to relatively manageable levels. As far as political elections went, I could not recall any where the global and personal stakes felt so high.

Joe Biden's election prompted a huge surge of relief across the climate research community. With his clear views on the urgency of tackling climate change, the prospects for progress at the upcoming talks in Glasgow had just got a whole lot brighter. And after seeing the ugly spectacle of supporters of the outgoing president invading the US Capitol, I watched the presidential inauguration with grateful fascination. Not only did Biden's speech refer to a cry for survival coming from the planet, he referred to the need to 'defend the truth' and to defeat the 'lies told for power and for profit'.[32] It was very welcome news that once again an American president was prepared to accept the scientific evidence on climate change and take action in response.

It was also a great relief that governments in the UK were mandated to take account of the scientific evidence thanks to the Climate Change Act. As a result, the country now had a clear target that was in line with the Paris Agreement, to reach net zero greenhouse gas emissions by 2050. Much more was needed to be done to get there. But a positive step had been taken in copying Ireland and setting up a citizens' assembly to consider how to do so, the Climate Assembly UK.[33] And in the run-up to the UK hosting the most important climate meeting since Paris, there were other encouraging developments.

The media landscape was changing dramatically. In February 2021, the *Daily Express*, the same newspaper that had printed the front page headline '100 reasons why global warming is natural' in the wake of Climategate, launched a 'crusade to save our environment' under the headline 'Join Our Green Britain Revolution'.[34] Other newspapers also ramped up their environmental coverage. Some like the *Independent* were known for their long-standing interest in green issues. Others like the *Sun*, who announced they were forming a 'green team' on the road to COP26, were more of a surprise.[35]

The world of business was also changing fast. Many companies announced plans to become carbon-neutral, including Microsoft, Sky and BP who formed part of a group of over 850 global corporations who had pledged to follow science-based targets consistent with the

Paris goals.[36] Climate deniers had claimed to be speaking up for private enterprise. Now private enterprise was speaking up for climate action.

Politicians, the media, business; at long last all were on board with addressing the climate crisis. The climate deniers had become an irrelevance; their claims that global warming had stopped or that extreme weather wasn't changing were widely seen as risible. For the first time in my twenty-five years of climate research, stopping the relentless march of global warming before it was too late began to look like a realistic prospect. Our science was being widely accepted and taken into account. I dared to hope that the battle against climate change denial might, finally, have been won.

Epilogue

The climate deniers have spouted hot air for years. Despite their dangerous lies that greenhouse gas emissions could do no harm, the world now finds itself in serious trouble, needing to make changes fast if we are going to save our habitable planet before it is too late. Just in time, governments have committed to doing what it takes to prevent global warming reaching unmanageable levels. Commitments are being made, not just by national parliaments but by industrial boards, institutions and local councils. A green recovery from Covid-19 offers opportunities to invest in more environmentally sustainable forms of energy, transport and food production. If the will is there to build a cleaner, healthier home on earth, the ways and means exist for us to do so.

Since I joined the hunt for climate fingerprints in 1996, we have found more and more evidence of the dominant effects of greenhouse gas emissions on warming temperatures, melting ice, raging storms and rising seas. In 2007, I helped present findings of the Intergovernmental Panel on Climate Change (IPCC). Not only was warming 'unequivocal', we showed that if warming continued, the risks to life on earth would become catastrophic with widespread extinctions, food shortages and coastal inundation. In 2008, the UK passed the Climate Change Act, mandating domestic targets for reducing greenhouse gas emissions as advised by the independent Climate Change Committee, and Barack Obama was elected to be the forty-fourth president of the United States of America campaigning for greenhouse gas regulation. Government leaders gathered in Copenhagen in December 2009 to agree international targets for reductions in emissions across the globe.

All this time, the forces of climate change denial were busy. The Global Climate Coalition, the Scientific Alliance, the George C. Marshall Institute, the Global Warming Policy Foundation, the Heartland Institute: all these and more, well funded by fossil fuel industries and supported by scientifically trained, self-styled 'climate sceptics', spread disinformation and confusion in the cause of climate delay. I saw with dismay how they attacked Ben Santer in 1996, falsely accusing him of distorting the IPCC conclusions. I was horrified by the antics of climate deniers in 2004 when they rushed to Moscow to support Russia's attempt to forestall ratification of the Kyoto Protocol. In 2009, in the lead-up to the vital meeting in Copenhagen, it was infuriating that false claims that scientists hid a decline in global temperatures gained widespread traction internationally.

In the aftermath of the collapse of the Copenhagen talks, the level of public engagement in the climate crisis reached a new low. Climate scientists needed to fight back. I helped launch that fightback when I presented a review of 110 peer-reviewed studies to a crowded room of journalists eager to know how I could possibly justify my conclusion that the evidence for human-induced climate change was even clearer than before. Over the next few years I worked hard on the next IPCC report and helped present our conclusions, the strongest yet, to governments in Stockholm. Humanity had already burned through most of the carbon from oil, gas and coal that could be permitted, the report showed, for global warming to be kept from reaching catastrophic levels. Scientists had given politicians the evidence they needed to act.

Two years later in Paris, governments agreed to do what it took to keep global warming to well below 2 degrees Celsius relative to pre-industrial levels and to pursue efforts to keep warming to below 1.5 degrees. With global warming already at more than 1 degree, this is a tough ask. But the high level of ambition, driven by the sobering findings of science, has now been enshrined in an international agreement. It is only a start and the aims of Paris still need to be translated into reality. But at long last the necessary global framework to do so has been put in place.

There is no time to lose. The global toll from floods, droughts and heatwaves continues to rise at a startling rate, their increasing intensity attributable, our research shows, to human-induced climate change. The people affected don't find our science surprising. They know that most of these disasters are not natural. Instead they see them rocket-fuelled by the human folly of unchecked emissions of greenhouse gases.

Now many more people are eager to engage in the pressing global issue of the climate crisis. In 2017, I met the randomly selected cross-section of Irish society making up their Citizens' Assembly as they discussed how to make their country a leader in addressing climate change. The experience showed me how ordinary people have an extraordinary capacity to engage constructively with the detailed evidence of science. The following year, Pierrette and I instigated a project called Climate Stories in which we brought together colleagues from the Met Office and the University of Exeter with local community groups and artists, writers and musicians to think creatively and imaginatively about tackling the climate crisis.[1] Nobody we met wanted to debate whether scientists had got it wrong about global warming. Instead, everybody wanted to imagine what we could achieve together – citizens, artists and scientists – to build a more sustainable future. And later that year, Greta Thunberg began her school strike for climate, starting a movement that has galvanized the world's youth who demand that politicians protect the future of this younger generation.

Thunberg's message to governments is that they should listen to the science. And hearteningly, that is now happening worldwide.

In 2018, climate science delivered its most urgent message yet in the form of the IPCC's Special Report on Global Warming of 1.5°C. To keep warming to 1.5 degrees, carbon dioxide emissions have to decline by about 45 per cent from their 2010 levels by 2030 and reach net zero by 2050.[2] Net zero means that by the middle of this century, humanity must draw down from the atmosphere as much carbon dioxide as it puts into it. It is challenging but not impossible. Net zero is a target that everyone can work towards.

In 2019, the UK government legislated that the country must reach net zero greenhouse gas emissions by 2050.[3] This can be done, the Climate Change Committee says, by a combination of measures that include expansion of electricity generation by offshore wind power and other renewable sources, decarbonizing the heating of buildings, the development of technologies to capture carbon emissions and store them underground, changes in farming practices, and electrification of transport with a phase-out of petrol and diesel cars.[4] It won't be easy but it can be achieved at a manageable cost of less than 1 per cent of GDP. There will be many benefits. Not only will the escalating economic and human costs of unchecked climate change be avoided, moving to net zero will mean cleaner air, healthier diets and quieter streets.

Thankfully, the momentum is now growing in favour of sustainable development that will save us from planetary catastrophe. Many companies have plans to become carbon-neutral, car manufacturers are switching away from petrol and diesel, and newspapers once hostile to climate science are now promoting a green agenda. And if the arrival of a global pandemic in 2020 has given humanity another colossal problem to grapple with, it also points the way to how we can build back economies better.

For the UK, the Climate Change Committee's 1,000-page report of December 2020 provides a blueprint for how to do so.[5] Large investments have been made in combating Covid-19 and as we rebuild our economy, suitable investment from both the public and private sector will be needed for the 2020s to be a decade of positive transformation. Cars and vans will have to switch to being electric, all new homes will need to be zero carbon and existing gas boilers will have to be replaced by hydrogen boilers or heat pumps. Capitalizing on how we have adapted to the virus, many office workers will need to take advantage of remote working for some of the week and more people will have to travel to work by walking, cycling or other zero-carbon transport. Innovative technologies will need to be developed to store and then re-use hydrogen power generated by renewable energy and also to draw down and capture any remaining carbon emissions and store them safely underground. Diets will need to change with reductions in the consumption of meat and dairy. None of this is pie in the sky according to the Climate Change Committee, whose members have expertise in engineering, behavioural science, economics and climate science.[6] All of it is doable, providing there is the will to do so.

That means the will of government to enact the right policies. It also means the will of individuals to do what we can. At home, despite the difficulties I recounted to the Citizens' Assembly in Ireland, I did eventually manage to have solar panels installed and the knowledge I am now saving money while doing a little bit for the planet is extremely satisfying. When my offices at the Met Office and the University of Exeter are open again, I will look forward to cycling there once more and getting some fresh air and exercise into the bargain. It's unlikely I'll be flying to huge international conferences again, like the American Geophysical Union meeting I attended in December 1996 that was my initiation into the charged nature of climate change science and the attacks by the climate deniers. But since the pandemic began, researchers have been developing new innovative ways of running conferences online that are just as productive and interesting as the old style physical meetings. I expect I'll go to plenty of them.

Many people would benefit from the transformational changes needed to take us to a net zero world. As the Committee on Climate

Change has pointed out, hundreds of thousands of jobs would be created in hydrogen production, in developing technologies for capturing carbon and in renovating and decarbonizing the nation's homes.[7] And absolutist remedies are not required. The road map set out by the Committee on Climate Change does not mean we all have to become vegans or give up on ever hugging loved ones who happen to live far away on other continents. Humanity will need to find new ways to draw down the residual emissions of carbon dioxide that our activities will continue to produce, even with successful efforts to dramatically reduce those emissions globally. But humankind has a remarkable ability to be inventive when there is the need to do so. Getting to a net zero world in time to prevent catastrophic climate change is both realistic and achievable.

There is one crucial caveat. It's related to why I got so emotional at the Berkeley Lab in California in March 2017 when I heard that incoming President Trump's appointee to head the US Environmental Protection Agency claimed that carbon dioxide is not a primary contributor to climate change. Avoiding catastrophic climate change is only realistic and achievable if the political will is there to listen to the experts and take heed of their advice. This is the process that the UK has enshrined in law through the Climate Change Act that commits governments to acting on the advice of the independent Climate Change Committee.

But the climate crisis cannot be solved by any one country alone. International action is required to drive down greenhouse gas emissions collectively. This is why Trump's withdrawal from the Paris Agreement because he was not prepared to believe the experts was so distressing. It is why so many climate scientists fervently hoped in November 2020 that he would not be elected for four more years. We value the defence of truth that President Biden referred to in his inauguration address. To save the planet, we have to defeat the lies told for power and for profit. For the safety of humanity on earth, we can't afford to have any more powerful leaders who promote the false belief of climate change denial.

Acknowledgements

This book has been twenty-five years in the making, starting with the day in July 1996 I began work at the Met Office Hadley Centre. I didn't know then that I would write a book about my experiences but I did realize very quickly that the Met Office was a great place to carry out research that was both intellectually fascinating and societally important, attributes of science at the Met Office that remain as true today as they were then. Many thanks to John Mitchell for recruiting me and to Simon Tett and Jonathan Gregory as well as Myles Allen at Oxford for showing me the ropes of climate research in those early days. Heartfelt thanks too to the many colleagues I've had the privilege to work with since, including the members of my research group – Claire Burke, Nikos Christidis, Andrew Ciavarella, Daniel Cotterill, Gareth Jones and Fraser Lott – and the many scientists at other institutes worldwide who've inspired and encouraged me over the years, including Gabi Hegerl, David Karoly, Ben Santer, Susan Solomon, Dáithí Stone, Michael Wehner and Francis Zwiers. Thank you to Nathan Bindoff and Pierre Friedlingstein in particular for advice on XBTs and TCRE (climate scientists love acronyms). And thank you to Sam Dean for inviting me to Wellington several times – to work together, to spend time with his wonderful family and explore his fabulous country – visits which have provided welcome respite from the stresses and strains of battling against climate change denial. Climate science is a global enterprise and I am grateful to the researchers from many different countries with whom I've had the good fortune to publish scientific papers and develop scientific assessments.

In recent years, thanks in part to Peter Cox who originally suggested the idea, I've been lucky to have a one day a week professorial position at the University of Exeter. This has enabled me to pursue new avenues of enquiry including the Climate Stories project that Pierrette Thomet and I instigated with funding from the Natural Environment Research Council. Collaborating with project team Miranda Addey, Stewart Barr, Rosie Eade, Sally Flint, Felicity Liggins, Fiona Lovell, Evelyn O'Malley, Dan Plews, Chris Rapley and Ewan Woodley, and twenty climate researchers from the University of Exeter and the Met Office, was an enriching and hopeful experience as artists, scientists and members of the public came together to think anew about the climate crisis. The experience was also helpful for me in developing my personal project of telling the story of how we showed that climate change was a human affair and of how climate deniers tried to stop that truth being widely known and acted upon by governments and citizens.

To help me tell that story I have had the good fortune to spend time at the Arvon Foundation's writing houses in the English countryside. I originally discovered the inspirational benefits of Arvon by spending a week at Totleigh Barton in 2009 under the remarkable tutelage of Mavis Cheek and Paul Sussman. For writing this story, the key impetus was an unforgettable course on popular science writing at The Hurst in 2016 under the equally remarkable tutelage of Aarathi Prasad and Michael Brooks. Thank you to Aarathi and Michael for their expert guidance that week, for Aarathi's reminders about the value of persistence and for Michael's advice and encouragement as I worked on the book in the years following the course. And thank you to Tania Hershman and Maria Fusco for a wonderfully stimulating week of hybrid writing at

Lumb Bank in 2019. Being among writers again was a big fillip.

Shortly after leaving Lumb Bank my book found an agent and then a publisher. Many thanks to Andrew Gordon at David Higham Associates and to Mike Harpley and James Pulford, my editor, at Atlantic, as well as to David Evans at DHA and Ian Greensill. It has been a great pleasure to work with such a team and know that the book has benefited greatly from this collaboration.

Finally, and most importantly of all, thank you to Pierrette for her love, support and wisdom during the writing of this book. She is my number one collaborator, as I am hers.

Endnotes

Preface

1. See https://www.ipcc.ch/report/sixth-assessment-report-working-group-i/
2. See Glasgow Climate Pact : https://unfccc.int/sites/default/files/resource/cop26_auv_2f_cover_decision.pdf
3. See https://www.metoffice.gov.uk/about-us/press-office/news/weather-and-climate/2021/2021-european-summer-temperature-impossible-without-climate-change
4. See https://news.un.org/en/story/2022/01/1110692
5. See https://www.theguardian.com/environment/2022/feb/08/its-all-a-bit-cynical-the-politicians-behind-the-tory-attack-on-net-zero-agenda
6. See https://www.thegwpf.org/content/uploads/2021/11/Steve-Koonin-2021-GWPF-Lecture.pdf

Prologue

1. Examples of publications based on research from the International Detection and Attribution Group at the time include: the annual report (that I co-edit), 'Explaining Extreme Events of 2016 From A Climate Perspective', published in the *Bulletin of the American Meteorological Society* which found that record heat in 2016 globally, over Asia and in the Bering Sea was attributable to human-induced climate change (https://journals.ametsoc.org/view/journals/bams/99/1/bams-explainingextremeevents2016.1.xml); the paper in *Nature* by Patricola and Wehner (Patricola, C.M., M.F. Wehner, 2018: 'Anthropogenic influences on major tropical cyclone events', *Nature*, Vol. 563, pp. 339–46), which found that climate change made the rainfall in Hurricanes Katrina, Irma and Maria more extreme; and the paper by Mueller et al. (Mueller, B.L., N.P. Gillett, A.H. Monahan, F.W. Zwiers, 2018: 'Attribution of Arctic sea ice decline from 1953 to 2012 to influences from natural, greenhouse gas, and anthropogenic forcing', *Journal of Climate*, Vol. 31, pp. 7771–87), which detected the fingerprint of greenhouse gas induced decline in Arctic sea ice.
2. See https://en.wikipedia.org/wiki/Scott_Pruitt
3. https://www.sciencemag.org/news/2017/03/epa-chief-says-carbon-

dioxide-not-primary-contributor-climate-change

4. In response to a question during the Senate Confirmation Hearing for the new administrator of the Environmental Protection Agency, Scott Pruitt stated that there had been a 'levelling off of warming' over the past two decades: https://www.epw.senate.gov/public/_cache/files/6d95005c-bd1a-4779-af7e-be831db6866a/scott-pruitt-qfr-responses-01.18.2017.pdf; https://www.nrdc.org/experts/david-doniger/scott-pruitts-climate-denial-shines-thru-his-senate-answers

5. The International Detection and Attribution Group is an informally structured group of scientists researching the detection and attribution of climate change who meet at a workshop on an annual basis. Some information about the group is provided at http://www.image.ucar.edu/idag/

6. These simulations were published in Stott, P.A., S.F.B. Tett, G.S. Jones, M.R. Allen, J.F.B. Mitchell, G.J. Jenkins, 2000: 'External control of 20th century temperature by natural and anthropogenic forcings', *Science*, Vol. 290, pp. 2133–37. The finding on the European heatwave of 2003 was published in Stott, P.A., D.A. Stone, M.R. Allen, 2004: 'Human contribution to the European heatwave of 2003', *Nature*, Vol. 432, pp. 610–14.

7. This analysis was subsequently published as Santer B.D., S. Solomon, F.J. Wentz, Q. Fu, S. Po-Chedley, C. Mears, J.F. Painter, C. Bonfils, 2017: 'Tropospheric warming over the past two decades', *Scientific Reports*, Vol. 7, p. 2336. In that analysis, calculations were made of temperature trends over a succession of twenty-year periods and over the full thirty-eight-year period (to date) of satellite measurements of atmospheric temperatures (1979–2016). Two different versions of three data sets of atmospheric temperatures measured from satellites were considered. For all of them, the most recent twenty-year period (1997–2016) of warming at that time was significantly larger, at the 10 per cent level or better, than estimates of twenty-year trends arising from natural internal variability. Over the full thirty-eight-year period (1979–2016), warming trends in the latest versions of the satellite data sets were all found to be more than five standard deviations removed from the mean of the distribution of natural internal variability of thirty-eight-year trends. Being greater than five standard deviations (what is known as five sigma) from the expectation of the null hypothesis (in this case that natural internal variability explains the observed trends) is a standard measure used in physics to detect fundamental particles such as the Higgs boson. Because the probability of a five-sigma excursion is vanishingly small – it is a 1 in 3.5 million chance – experimentalists at CERN could announce their discovery of the Higgs boson when they could rule out the null hypothesis of chance fluctuations producing their measurements at this five-sigma level. Likewise, Ben Santer showed that the null hypothesis of chance fluctuations producing the observed temperature trends seen from satellites could be ruled out at this same five-sigma level. This confirms that global warming was detected in satellite data to the same level of confidence that the Higgs boson was detected from data collected at CERN. For explanation of the significance of five sigma for physics see: https://blogs.scientificamerican.com/observations/five-sigmawhats-that/

8. For the impact on Lakeland sheep farmers, see https://www.theguardian.

com/uk/2009/dec/29/sheep-farmers-
chernobyl-meat-restricted

9. https://uk-air.defra.gov.uk/research/
ozone-uv/moreinfo?view=recovery

10. A perspective on the history of the Met
Office Hadley Centre (subsequently
renamed the Hadley Centre for

Climate Science and Services) in its
thirtieth anniversary year in 2020 is
provided at https://www.theguardian.
com/environment/2020/may/27/
hadley-climate-centre-turns-30-the-
human-fingerprint-is-everywhere

11 https://www.ipcc.ch/report/ar4/wg1/

1 Fingerprinting the climate

1. The research I presented at the
1996 fall meeting of the American
Geophysical Union was further
developed and subsequently published
as Stott, P.A., S.F.B. Tett, 1998: 'Scale
dependent detection of climate change',
Journal of Climate, Vol. 11, pp. 3282–
94.

2. https://www.agu.org

3. The IPCC's Second Assessment Report
on the Science of Climate Change:
https://www.ipcc.ch/report/ar2/wg1/

4. The results I presented (and
subsequently published as Stott, P.A.,
S.F.B. Tett, 1998: 'Scale dependent
detection of climate change', *Journal
of Climate*, Vol. 11, pp. 3282–94) were
consistent with a paper published the
year before by Ben Santer (Santer,
B.D., K.E. Taylor, T.M.L. Wigley,
J.E. Penner, P.D. Jones, U. Cubasch,
1995: 'Towards the detection and
attribution of an anthropogenic effect
on climate', *Climate Dynamics*, Vol. 12,
pp. 77–100), which found that global
mean temperature trends are outside
the envelope of natural internal
variability over twenty years or longer
but not over ten years.

5. This work was published in Tett, S.F.B.,
J.F.B. Mitchell, D.E. Parker, M.R.
Allen, 1996: 'Human influence on
the atmospheric vertical temperature
structure: detection and observations',
Science, Vol. 274, pp. 1170–73.

6. The research Myles presented was
further developed and subsequently

published as Allen, M.R., S.F.B. Tett,
1999: 'Checking for model consistency
in optimal fingerprinting', *Climate
Dynamics*, Vol. 15, pp. 419–34.

7. Michaels' position on global warming
was set out in his 1992 book *Sound and
Fury: The Science and Politics of Global
Warming* (Cato Institute, Washington
DC).

8. The argument that Michaels made at
the American Geophysical Union was
published in a short letter in *Nature*
as Michaels, P.J., P.C. Knappenberger,
1996: *Nature*, Vol. 384, pp. 522–23.
In it, he claimed that Ben Santer's
finding of human influence on climate
published in *Nature* (Santer, B.D.,
K.E. Taylor, T.M.L. Wigley, P.D. Jones,
D.J. Karoly, J.F.B. Mitchell, A.H.
Oort, J.E. Penner, V. Ramaswamy,
M.D. Schwarzkopf, R.S. Stouffer,
S.F.B. Tett, 1996: 'A search for human
influences on the thermal structure
in the atmosphere', *Nature*, Vol. 382,
pp. 39–46) was based on a pattern
correlation that was a manifestation
of natural variability (a thesis that was
also proposed in an accompanying
letter to *Nature* by G.R. Weber, 1996:
Nature, Vol. 382, pp. 523–24). This
proposition was rebutted in a response
by Santer and colleagues also published
in the same edition of *Nature* (Santer,
B.D. et al., 1996: *Nature*, Vol. 382,
p. 524) where they pointed out that the
warming of the southern hemisphere
relative to the northern hemisphere

was entirely consistent with the expectations of anthropogenic forcing on the climate.

9. The attacks on Ben Santer, the defence from the scientific community and the position of Patrick Michaels are described in a series of letters published in the *Bulletin of the American Meteorological Society*, Vol. 77, pp. 1961–66 and Vol. 78, pp. 81–83. Michaels and co-authors sum up with: 'The real issue then is the political misuse of the IPCC report and of climate science.' The reason why these claims are false that Ben Santer distorted the IPCC report for political ends are discussed in Chapter 2 and are analysed in *Merchants of Doubt* by Naomi Oreskes and Erik M. Conway (Bloomsbury).

10. https://www.margaretthatcher.org/document/107346; *In the Eye of the Storm* by John Houghton (Lion Hudson).

11. https://unfccc.int/process-and-meetings/the-convention/what-is-the-united-nations-framework-convention-on-climate-change

12. https://en.wikipedia.org/wiki/Global_Climate_Coalition; *The Rough Guide to Climate Change* by Robert Henson (Rough Guides).

13. The research Ben presented at the American Geophysical Union was published in *Nature* on 4 July 1996 (Santer, B.D., K.E. Taylor, T.M.L. Wigley, P.D. Jones, D.J. Karoly, J.F.B. Mitchell, A.H. Oort, J.E. Penner, V. Ramaswamy, M.D. Schwarzkopf, R.S. Stouffer, S.F.B. Tett, 1996: 'A search for human influences on the thermal structure in the atmosphere', *Nature*, Vol. 382, pp. 39–46). The coloured map of observed temperature changes is Figure 1j of that paper based on radiosonde-based temperature measurements from 1963 to 1988. This illustration of observed temperature changes also appears in the IPCC

Working Group I Contribution to the Second Assessment Report as Figure 8.7c in Chapter 8 (see https://www.ipcc.ch/report/ar2/wg1/).

14. Plots showing how the different factors contribute to atmospheric temperature changes also appear in Figure 1 of the Santer et al. *Nature* paper. The effects of increases in greenhouse gases on atmospheric temperatures are shown in Figure 8.7a of the IPCC Working Group I Contribution to the Second Assessment Report and the combined effects of greenhouse gas increases and changes in aerosols are shown in Figure 8.7b (see https://www.ipcc.ch/report/ar2/wg1/). The state of knowledge at the time of the 1996 American Geophysical Union fall meeting on this topic is summarized in the Summary of Chapter 8 of the report – 'Pattern correspondences using combined CO2 and aerosol signals are generally higher than those obtained if model predictions are based on CO2 alone. Furthermore, the probability is very low that these correspondences could occur by chance as a result of natural internal variability. The vertical patterns of change are also inconsistent with the response patterns expected for solar and volcanic forcing. Increasing confidence in the emerging identification of a human-induced effect on climate comes primarily from such pattern-based work.'

15. Mitchell, J.F.B., T.C. Johns, J.M. Gregory, S.F.B. Tett, 1995: 'Climate response to increasing levels of greenhouse gases and sulphate aerosols', *Nature*, Vol. 376, pp. 501–04.

16. The original manuscript of my paper – Stott, P.A., S.F.B. Tett, 1998: 'Scale dependent detection of climate change', *Journal of Climate*, Vol. 11, pp. 3282–94 – was received by the *Journal of Climate* on 7 July 1997, was returned

in revised form following anonymous peer review on 4 February 1998 and was published in the December 1998 issue of the *Journal of Climate*.

17. *The Rough Guide to Climate Change* by Robert Henson (Rough Guides); https://www.easterbrook.ca/steve/2015/08/who-first-coined-the-term-greenhouse-effect/

18. The work of Eunice Foote has been forgotten until relatively recently; in 2016, climate scientist and communicator Katharine Hayhoe found Foote's contribution after a colleague asked her why there were no women in the history of the discipline. This rediscovery is recounted at https://www.climatechangenews.com/2016/09/02/the-woman-who-identified-the-greenhouse-effect-years-before-tyndall/

19. See Jackson, R., 2019: 'Eunice Foote, John Tyndall and a question of priority', *Notes and Records of the Royal Society*, Vol. 74, pp. 105–18, doi.org/10.1098/rsnr.2018.0066.

20. A calculation of the warming effect provided by the earth's natural greenhouse effect is provided in the IPCC's Working Group I Contribution to the Fourth Assessment Report in the Frequently Asked Question 1.1: What Factors Determine Earth's Climate? (https://www.ipcc.ch/site/assets/uploads/2018/03/ar4-wg1-chapter1.pdf). The energy from the sun that is not reflected back to space is absorbed by the earth's surface and atmosphere. To balance the incoming energy, the earth must radiate the same amount of energy back to space by emitting longwave radiation. Without the greenhouse effect this would mean the surface would have to have a temperature of around −19°C which is 33 degrees colder than the actual global mean surface temperature of about 14°C. The reason the earth's surface is so much warmer is due to the presence of greenhouse gases in the atmosphere which act as a partial blanket for the longwave radiation coming from the earth's surface.

21. See *The Rough Guide to Climate Change* by Robert Henson (Rough Guides), and the Timeline of Climate Modelling provided by Carbon Brief at: https://www.carbonbrief.org/timeline-history-climate-modelling

22. See http://sciencenordic.com/vikings-grew-barley-greenland for information on Vikings and https://www.museumoflondon.org.uk/discover/frost-fairs for information on London's Frost Fairs. While such instances are evidence of local climate being different than present, such regional perturbations are not necessarily associated with coherent warming or cooling globally. Also, the viability of Frost Fairs was as much associated with the configuration of the river at that time as the local climate; see http://www.realclimate.org/index.php/archives/2004/11/little-ice-age-lia/ for a discussion of the so-called Little Ice Age.

23. The 'Keeling curve' of atmospheric carbon dioxide concentrations can be explored at https://scripps.ucsd.edu/programs/keelingcurve/.

24. Carbon Brief has published an analysis by John Mitchell of the seminal paper by Manabe and Wetherald (Manabe, S., R.T. Wetherald, 1967: *Journal of the Atmospheric Sciences*, Vol. 24, pp. 241–59) at https://www.carbonbrief.org/prof-john-mitchell-how-a-1967-study-greatly-influenced-climate-change-science

25. Carbon dioxide concentrations can be compared with the record of global average temperatures and other climate indicators at https://www.metoffice.gov.uk/hadobs/monitoring/dashboard.html

26. See https://www.nytimes.com/1988/06/24/us/global-warming-has-begun-expert-tells-senate.html

27. See *Merchants of Doubt* by Naomi Oreskes and Erik M. Conway (Bloomsbury) and *Losing Earth* by Nathanial Rich (Picador).

28. Klaus Hasselmann was founding director of the Max Planck Institute of Meteorology, Hamburg from February 1975 to November 1999; see https://en.wikipedia.org/wiki/Klaus_Hasselmann. He is probably best known in climate science for the Hasselmann model of climate variability where the ocean (which has a long memory) integrates stochastic forcing, transforming white noise into red noise and explaining the red-noise signals seen in the climate. Hasselmann's many publications are listed at https://www.mpimet.mpg.de/en/staff/externalmembers/klaus-hasselmann/publications/

29. This was published as Hasselmann, K., 1979: 'On the signal to noise problem in atmospheric response studies', in: *Meteorology of Tropical Oceans*, D.B. Shaw (ed.), Royal Meteorological Society London, pp. 251–79. The seminal nature of this 1979 paper is assessed in a review paper by Ben Santer and co-authors published in 2019 in the journal *Nature Climate Change* (Santer B.D., C.J.W. Bonfils, Q. Fu, J.C. Fyfe, G.C. Hegerl, C. Mears, J.F. Painter, S. Po-Chedley, F.J. Wentz, M.D. Zelinka, C.-Z. Zhou, 2019: 'Celebrating the anniversary of three key events in climate change science', *Nature Climate Change*, Vol. 9, pp. 180–82.) In that review paper, Santer et al. point out that 'One key insight in Hasselmann's 1979 paper was that analysts should look at the statistical significance of global geographical patterns of change. Previous work had assessed the significance of the local climate response to a particular external forcing at thousands of individual model grid-points. Climate information at these individual locations was correlated in space and in time, hampering assessment of overall significance... Instead of looking for a needle in a tiny corner of a large haystack (and then proceeding to search the next tiny corner), Hasselmann advocated for a more efficient strategy – searching the entire haystack simultaneously.'

30. A primer on climate modelling has been published by Carbon Brief at https://www.carbonbrief.org/qa-how-do-climate-models-work

31. The progress Ben made in detecting surface temperature changes at that time can be charted through a number of important contributions to the climate science literature, including Santer, B.D., W. Brüggemann, U. Cubasch, K. Hasselmann, H. Höck, E. Maier-Reimer, U. Mikolajewicz, 1994: 'Signal-to-noise analysis of time-dependent greenhouse warming experiments, Part 1: Pattern analysis', *Climate Dynamics*, Vol. 9, pp. 267–85; Santer, B.D., U. Mikolajewicz, W. Brüggemann, U. Cubasch, K. Hasselmann, H. Höck, E. Maier-Reimer, T.M.L. Wigley, 1995: 'Ocean variability and its influence on the detectability of greenhouse warming signals', *Journal of Geophysical Research*, Vol. 100, No. C6, pp. 10693–725; and Santer B.D., K.E. Taylor, T.M.L. Wigley, J.E. Penner, P.D. Jones, U. Cubasch, 1995: 'Towards the detection and attribution of an anthropogenic effect on climate', *Climate Dynamics*, Vol. 12, pp. 77–100.

32. Professor Gabriele Hegerl, now a Fellow of the Royal Society, is currently based at the University of Edinburgh (https://blogs.ed.ac.uk/ghegerl/).

33. For information about climate science at the Lawrence Livermore National Laboratory, see https://climate.llnl.gov

34. The physics of this decrease in temperature with height – the so-called lapse rate of the atmosphere – is described in more detail in *Global Warming, The Complete Briefing* by John Houghton (Cambridge University Press, fifth edition, p. 21) and is derived mathematically in his scholarly account of basic meteorology, *The Physics of Atmospheres* by John Houghton (Cambridge University Press).

35. The layers of varnish analogy is taken from Bill Bryson's book *A Short History of Nearly Everything* by Bill Bryson (Puffin), http://www.fogonazos. es/2007/08/bill-brysons-top-10-science-quotes.html

36. See, for example, Figure 5.1 of *The Physics of Atmospheres* by John Houghton (Cambridge University Press, third edition).

37. The explanation for why the stratosphere cools with increasing greenhouse gas concentrations is relatively complex but involves the fact that increased atmospheric greenhouse gases in the troposphere reduce the amount of heat radiated into the stratosphere. See https://www. yaleclimateconnections.org/2013/09/ vertical-human-fingerprint-found-in-stratospheric-cooling-tropospheric-warming/

38. Karoly, D.J., J.A. Cohen, G.A. Meehl, J.F.B. Mitchell, A.H. Oort, R.J. Stouffer, R.T. Wetherald, 1994: 'An example of fingerprint detection of greenhouse climate change', *Climate Dynamics*, Vol. 10, pp. 97–105; Mitchell, J.F.B., T.C. Johns, J.M. Gregory, S.F.B. Tett, 1995: 'Climate response to increasing levels of greenhouse gases and sulphate aerosols', *Nature*, Vol. 376, pp. 501–04.

39. The two seminal papers published in 1996 by Gabi Hegerl and Ben Santer are respectively: Hegerl, G.C., H. von Storch, K. Hasselmann, B.D. Santer, U. Cubasch, P.D. Jones, 1996: 'Detecting greenhouse gas induced climate change with an optimal fingerprint method', *Journal of Climate*, Vol. 9, pp. 2281–306; and Santer, B.D., K.E. Taylor, T.M.L. Wigley, P.D. Jones, D.J. Karoly, J.F.B. Mitchell, A.H. Oort, J.E. Penner, V. Ramaswamy, M.D. Schwarzkopf, R.S. Stouffer, S.F.B. Tett, 1996: 'A search for human influences on the thermal structure in the atmosphere', *Nature*, Vol. 382, pp. 39–46. The full context of how the results of these two papers fit into the wider picture of research findings available at the time of the IPCC Second Assessment Report is provided in Chapter 8 of the report (https://www.ipcc.ch/report/ar2/wg1/). As the summary of Chapter 8 states: 'To better address the attribution problem, a number of recent studies have compared observations with the spatial patterns of temperature change predicted by models in response to anthropogenic forcing... These comparisons have been made both at the earth's surface and in vertical section through the atmosphere... The pattern correspondences increase with time, as one would expect as an anthropogenic signal increases in strength... Furthermore, the probability is very low that these correspondences could occur by chance as a result of natural internal variability. The vertical patterns of change are also inconsistent with the response patterns expected for solar and volcanic forcing. Increasing confidence in the emerging identification of a human-induced effect on climate comes primarily from pattern-based work.' Figures taken from Hegerl et al. and Santer et al. appear in Chapter 8 as Figure 8.5 and Figure 8.7 respectively.

40. Accounts of this meeting are provided in *In the Eye of the Storm* by John Houghton (Lion Hudson), *Merchants of Doubt* by Naomi Oreskes and Erik

M. Conway (Bloomsbury), and in Houghton, J., 2008: 'Madrid 1995: Diagnosing Climate Change', *Nature*, Vol. 455, pp. 737–38, https://www.nature.com/articles/455737a.pdf

2 Confronted by denial

1. A comprehensive account of the Kyoto Protocol and the negotiations that took place in Kyoto in December 1997 is provided in *The Kyoto Protocol: A Guide and Assessment* by Michael Grubb with Christiaan Vrolijk and Duncan Brack (Royal Institute of International Affairs).

2. The results we presented are provided in the report 'Climate Change and its impacts: a global perspective. Some recent results from the UK research programme', December 1997, http://cedadocs.ceda.ac.uk/237/1/COP3.pdf

3. For a detailed perspective on the political and legal foundations of the Kyoto Protocol, see Chapter 2 of *The Kyoto Protocol: A Guide and Assessment* by Michael Grubb with Christiaan Vrolijk and Duncan Brack (Royal Institute of International Affairs).

4. See Chapters 3 and 4 of *The Kyoto Protocol: A Guide and Assessment by Michael Grubb* with Christiaan Vrolijk and Duncan Brack (Royal Institute of International Affairs).

5. *Merchants of Doubt* by Naomi Oreskes and Erik M. Conway (Bloomsbury), p. 130.

6. https://en.wikipedia.org/wiki/Sun_Myung_Moon

7. *Merchants of Doubt* by Naomi Oreskes and Erik M. Conway (Bloomsbury), p. 134.

8. *Merchants of Doubt* by Naomi Oreskes and Erik M. Conway (Bloomsbury), p. 208.

9. Singer, S. 1996, Letter to the Editor, *Wall Street Journal*, New York, 11 July. See also Edwards, P.N., S.H. Schneider, 1997: 'The 1995 IPCC Report: Broad Consensus or "Scientific Cleansing"?', *Ecofable/Ecoscience*, 1:1 (1997), pp. 3–9, http://www.pne.people.si.umich.edu/PDF/ecofables.pdf

10. *In the Eye of the Storm* by John Houghton (Lion Hudson), pp. 187–88; *Merchants of Doubt* by Naomi Oreskes and Erik M. Conway (Bloomsbury), pp. 208–09. The unedited version of the letter was subsequently published by the *Bulletin of the American Meteorological Society*, Vol. 77, pp. 1963–65.

11. *Merchants of Doubt* by Naomi Oreskes and Erik M. Conway (Bloomsbury), p. 210.

12. Ben's reaction to the use of the word 'cleansing' is reported at http://www.theguardian.com/environment/2010/feb/09/ipcc-report-author-data-openness

13. The email is available at http://www.realclimate.org/docs/Email_25_July_1996.pdf

14. See point 5 of Ben's email rebutting Singer's accusations at http://www.realclimate.org/docs/Email_25_July_1996.pdf

15. From Ben's email to colleagues: http://www.realclimate.org/docs/Email_25_July_1996.pdf

16. Section 3.2.4 of Working Group I Contribution to the IPCC's Second Assessment Report published 1996 (https://www.ipcc.ch/report/ar2/wg1/).

17. This fallacious argument used by climate deniers is dealt with at https://skepticalscience.com/ocean-and-global-warming.htm

18. *Merchants of Doubt* by Naomi Oreskes and Erik M. Conway (Bloomsbury), pp. 126–30 and pp. 132–35.

19. A description of the ozone hole and its causes is contained in *Global Warming, the Complete Briefing*, by John Houghton (Cambridge University Press, fifth edition), pp. 50–51.

20. See https://www.newscientist.com/article/mg20727771-400-zeros-to-heroes-how-we-almost-missed-the-ozone-hole/

21. See https://en.wikipedia.org/wiki/Susan_Solomon for a concise account of Susan Solomon's role (Ozone Hole section). Her own account of the discovery and explanation of the ozone hole is at https://www.nature.com/articles/d41586-019-02837-5 and the paper announcing her discovery is Solomon, S., R. Garcia, F. Rowland et al., 1986: 'On the depletion of Antarctic ozone', *Nature*, Vol. 321, pp. 755–58.

22. See *Merchants of Doubt* by Naomi Oreskes and Erik M. Conway (Bloomsbury), p. 203, p. 130 and p. 133; Patrick J. Michaels, 'Apocalypse machine blows up', *Washington Times*, 1 November 1991; Patrick Michaels, 'More hot air from the stratosphere', *Washington Times*, 27 October 1992; and S. Fred Singer, 'The Hole Truth

about CFCs', *Chemistry and Industry*, 21 March 1994: 240; S. Fred Singer, 'Bad Science Pulling the Plug on CFCs?', *Washington Times*, 22 February 1994.

23. For further discussion of the persistent dissemination of the false theory that the ozone hole was caused by volcanic eruptions and not CFCs and a detailed account of its easy rebuttal, see the President's Lecture given at the 159th annual meeting of the American Association for the Advancement of Science on 14 February 1993 by 1995 Nobel Prize Winner, F. Sherwood Rowland: https://science.sciencemag.org/content/sci/260/5114/1571.full.pdf

24. See World Meteorological Organization, Global Ozone Research and Monitoring Project – Report No. 58, 8 February 2019 at https://public.wmo.int/en/media/news/scientific-assessment-confirms-start-of-recovery-of-ozone-layer

25. These late-night negotiations together with their wider context are described in *The Kyoto Protocol: A Guide and Assessment* by Michael Grubb with Christiaan Vrolijk and Duncan Brack (Royal Institute of International Affairs), pp. 95–111.

3 Strengthening the science

1. This paper was eventually published as Tett S.F.B., P.A. Stott, M.R. Allen, W.J. Ingram, J.F.B. Mitchell, 1999: 'Causes of twentieth-century temperature change near the Earth's surface', *Nature*, Vol. 399, pp. 569–72.

2. This was published as Allen, M.R., S.F.B. Tett, 1999: 'Checking for model consistency in optimal fingerprinting', *Climate Dynamics*, Vol. 15, pp. 419–34.

3. See https://scied.ucar.edu/sunspot-cycle.

4. The three elements of attribution outlined here – estimating the linear combination of natural and anthropogenic factors explaining observed changes, taking account of too weak or too strong climate model responses to external forcings of climate, and checking the climate model's representation of internal variability – were all set out in Allen, M.R., S.F.B. Tett, 1999: 'Checking for model consistency in optimal

fingerprinting', *Climate Dynamics*, Vol. 15, pp. 419–34.

5. Both of the rival explanations for the early-century warming were shown in the two figures in the paper. The paper's abstract stated, 'We find that solar forcing may have contributed to the temperature changes early in the century, but anthropogenic causes combined with natural variability would also present a possible explanation.'

6. The statistics of *Nature* submissions are at https://www.nature.com/nature/for-authors/editorial-criteria-and-processes

7. The climate model we used for the 1999 Tett et al. *Nature* paper was HadCM2. The atmospheric component has 19 levels with a horizontal resolution of 2.5 degrees of latitude by 3.75 degrees of longitude, which produces a global grid of 96 x 73 grid cells. This is equivalent to a surface resolution of about 417 km x 278 km at the equator, reducing to 295 km x 278 km at 45 degrees of latitude. The oceanic component has 20 levels with a horizontal resolution of 2.5 degrees of latitude by 3.75 degrees of longitude. See https://www.metoffice.gov.uk/research/approach/modelling-systems/unified-model/climate-models/hadcm2. The requirement to parameterize sub-grid scale processes in climate models is explained at https://www.carbonbrief.org/qa-how-do-climate-models-work

8. The Carbon Brief primer on climate modelling includes information about flux correction: https://www.carbonbrief.org/qa-how-do-climate-models-work

9. *Weather Prediction by Numerical Process*, Lewis Fry Richardson (Cambridge Mathematical Library,

Cambridge University Press, second edition).

10. https://en.wikipedia.org/wiki/Lewis_Fry_Richardson.

11. See https://www.carbonbrief.org/timeline-history-climate-modelling.

12. Phillips, N.A., 'The general circulation of the atmosphere: a numerical experiment', 1956: *Quarterly Journal of the Royal Meteorological Society*, Vol. 82, pp. 123–64; Manabe, S., R.T. Wetherald, 1967: *Journal of the Atmospheric Sciences*, Vol. 24, pp. 241–59; Manabe, S., K. Bryan, M. Spelman, 'A Global Ocean-Atmosphere Climate Model. Part I: The Atmospheric Circulation', 1975: *Journal of Physical Oceanography*, Vol. 5, pp. 3–29.

13. The prediction was published as Hansen J., R. Ruedy, M. Sato, 'Potential climate impact of Mount Pinatubo eruption', 1992: *Geophysical Research Letters*, Vol. 19, pp. 215–18. For verification, see 'A Pinatubo Climate Modeling Investigation', a conference paper by Hansen, J. et al., https://www.researchgate.net/publication/256423431_A_Pinatubo_Climate_Modeling_Investigation

14. For the history of supercomputing at the Met Office see https://en.wikipedia.org/wiki/Met_Office#High_performance_computing.

15. See https://www.metoffice.gov.uk/research/approach/modelling-systems/unified-model/climate-models/hadcm3. In addition to the higher ocean resolution, the main differences from the previous model, HadCM2, are a more sophisticated radiation scheme, the inclusion of the direct impact of convection on momentum and the inclusion of a new land surface scheme that includes a better representation of evaporation, freezing and melting of soil moisture. Scientific references are Gordon, C., C. Cooper, C.A. Senior,

H. Banks, J.M. Gregory, T.C. Johns, J.F.B. Mitchell, R.A. Wood (2000), 'The simulation of SST, sea ice extents and ocean heat transports in a version of the Hadley Centre coupled model without flux adjustments', *Climate Dynamics*, Vol. 16, pp. 147–68; Pope, V.D., M.L. Gallani, P.R. Rowntree, R.A. Stratton, 2000: 'The impact of new physical parametrizations in the Hadley Centre climate model – HadAM3', *Climate Dynamics*, Vol. 16, pp. 123–46.

16. For an in-depth analysis of chaos and its implications for weather forecasting and climate prediction, see http://www.realclimate.org/index.php/archives/2005/11/chaos-and-climate/.

17. The final form of this graph appeared as Figure 1 of the published paper, in which an additional line was added showing a possible future trajectory of global warming if greenhouse gases were to follow one particular scenario of future emissions.

18. This appeared in revised form as Figure 4 of the Summary for Policymakers of Working Group I's Contribution to the IPCC's Third Assessment Report: https://www.ipcc.ch/site/assets/uploads/2018/07/WG1_TAR_SPM.pdf

19. The paper was published as Stott, P.A., S.F.B. Tett, G.S. Jones, M.R. Allen, J.F.B. Mitchell, G.J. Jenkins, 2000: 'External control of 20th century temperature by natural and anthropogenic forcings', *Science*, Vol. 290, pp. 2133–37.

20. Published accounts of the Shanghai meeting appear in *In the Eye of the Storm* by John Houghton (Lion Hudson) and *Merchants of Doubt* by Naomi Oreskes and Erik M. Conway (Bloomsbury).

21. A detailed account and analysis of the discussions during the Shanghai meeting is provided by A.C. Petersen, *Simulating Nature: A Philosophical Study of Computer-Simulation Uncertainties and Their Role in Climate Science and Policy Advice* (Het Spinhuis, Amsterdam).

22. In the second edition of his book, unlike the first edition, Petersen attributes statements to particular countries, in this case to Saudi Arabia. See page 161 of A.C. Petersen, *Simulating Nature: A Philosophical Study of Computer-Simulation Uncertainties and Their Role in Climate Science and Policy Advice* (second edition, CRC Press).

23. *In the Eye of the Storm* by John Houghton (Lion Hudson), p. 202.

24. Working Group I Contribution to the IPCC's Third Assessment Report is at https://www.ipcc.ch/report/ar3/wg1/.

25. See https://en.wikipedia.org/wiki/2000_United_States_presidential_election_recount_in_Florida#Palm_Beach_County's_butterfly_ballots

26. https://en.wikipedia.org/wiki/Kyoto_Protocol#Ratification_process

4 Ambushed by power

1. This was published in 2005 as The International Ad Hoc Detection and Attribution Group, 2005: 'Detecting and attributing external influences on the climate system: A review of recent advances', *Journal of Climate*, Vol. 18, pp. 1291–1314.

2. *The Russia Journal Daily*, 24 February 2004. See House of Lords report which refers to https://publications.

parliament.uk/pa/ld200304/ldselect/ldeucom/179/17909.htm. Illarionov was also reported in April as describing the Kyoto Protocol as 'an interstate Auschwitz'; see https://www.independent.co.uk/news/world/europe/kyoto-treaty-is-an-auschwitz-for-russia-says-putins-adviser-560033.html

3. Corbyn's Weather Action Network: http://www.weatheraction.com

4. For an assessment of the so-called iris hypothesis, see http://www.realclimate.org/index.php/archives/2015/04/the-return-of-the-iris-effect/

5. Document titled 'Moscow Academy of Sciences – UK Climate Change Spokespeople' provided in brown paper parcel at foyer of Russian Academy of Sciences, Private Archive.

6. Document titled 'Moscow Academy of Sciences – UK Climate Change Spokespeople' provided in brown paper parcel at foyer of Russian Academy of Sciences, Private Archive.

7. This is Figure 4 of the Summary for Policymakers of the Working Group I Contribution to the IPCC's Third Assessment Report: https://www.ipcc.ch/site/assets/uploads/2018/07/WG1_TAR_SPM.pdf

8. This is Figure 1b of the Summary for Policymakers of the Working Group I Contribution to the IPCC's Third Assessment Report: https://www.ipcc.ch/site/assets/uploads/2018/07/WG1_TAR_SPM.pdf

9. http://www.realclimate.org/index.php/archives/2004/11/hockey-stick/

10. Illarionov's assertions about climate science were also provided to the press conference he gave at the end of the meeting as reported at https://www.sysecol2.ethz.ch/pdfs/Illarionov_Interv._9.Jul.04.pdf

11. The original findings of Mann and colleagues which spanned the past 600 years were published in Mann, M.E., R.S. Bradley, M.K. Hughes, 1998: 'Global-scale temperature patterns and climate forcing over the past six centuries', *Nature*, Vol. 392, pp. 779–87. The analysis was then extended to the past millennium and published in Mann, M.E., R.S. Bradley, M.K. Hughes, 1999, 'Northern hemisphere temperatures during the past millennium: Inferences, uncertainties, and limitations', *Geophysical Research Letters*, Vol. 26, pp. 759–62.

12. As shown, for example, by Figure 1 of The International Ad Hoc Detection and Attribution Group, 2005: 'Detecting and attributing external influences on the climate system: A review of recent advances', *Journal of Climate*, Vol. 18, pp. 1291–1314.

13. *The Hockey Stick and the Climate Wars, Dispatches from the Front Lines* by Michael E. Mann (Columbia University Press), Chapter 8, pp. 108–25.

14. Funding received by Inhofe from oil and gas interests was reported to the European Parliament by Geoffrey Suppran at its Hearing on Climate Change Denial held on 21 March, 2019 (https://multimedia.europarl.europa.eu/en/committee-on-petitions-ordinary-meeting-ordinary-meeting_20190321-1030-COMMITTEE-PETI_vd) and by Naomi Oreskes in her testimony to the US Congress House Committee on Oversight and Reform, Subcommittee on Civil Rights and Civil Liberties on 23 October 2019 (https://congress.gov/116/meeting/house/110126/witnesses/HHRG-116-GO02-Wstate-OreskesN-20191023.pdf).

15. *The Hockey Stick and the Climate Wars, Dispatches from the Front Lines* by Michael E. Mann (Columbia University Press), Chapter 8, pp. 113–22.

16. Soon, W., S. Baliunas, 2003: 'Proxy climatic and environmental changes over the past 1000 years', *Climate*

Research, Vol. 23, pp. 89–110; Soon, W., S. Baliunas, C. Idso, D.R. Legates, 2003: 'Reconstructing climatic and environmental changes of the past 1000 years: A reappraisal *Energy and Environment*, Vol. 14, pp. 233–96.

17. *The Hockey Stick and the Climate Wars, Dispatches from the Front Lines* by Michael E. Mann (Columbia University Press), Chapter 8, p. 120.

18. *The Hockey Stick and the Climate Wars, Dispatches from the Front Lines* by Michael E. Mann (Columbia University Press), Chapter 8, p. 122.

19. McIntyre, S., R. McKitrick, 2003: 'Corrections to the Mann et al. [1998] Proxy Database and Northern Hemisphere Average Temperature Series', *Energy and Environment*, Vol. 14, pp. 751–77.

20. *The Hockey Stick and the Climate Wars, Dispatches from the Front Lines* by Michael E. Mann (Columbia University Press), Chapter 8, pp. 122–23.

21. *The Hockey Stick and the Climate Wars, Dispatches from the Front Lines* by Michael E. Mann (Columbia University Press), Chapter 8, p. 123.

22. See https://en.wikipedia.org/wiki/Climate_Stewardship_Acts

23. As presented in the slide set 'CLIMATE CHANGE *some aspects of consensus*', printed copies of which were provided in the brown paper parcels given to participants of the meeting in the foyer of the Russian Academy of Sciences.

24. Slide 4 of 'CLIMATE CHANGE *some aspects of consensus*' by Richard S. Lindzen.

25. Slide 4 of 'CLIMATE CHANGE *some aspects of consensus*' by Richard S. Lindzen states that 'a more accurate and relevant statement' of the IPCC conclusion that most of the warming over the last 50 years is likely to have been due to the increase in greenhouse gas concentrations is 'If these models are correct, then man has accounted for over four times the observed warming over the past century'. His slide then goes on to state: 'This statement illustrates that the observations do not support the likelihood of dangerous warming, but our ignorance may be sufficient to allow the possibility. In point of fact, our ignorance is not that great.'

26. Slide 12 of 'CLIMATE CHANGE *some aspects of consensus*' by Richard S. Lindzen states: 'Sir John Houghton made the casual claim that a warmer world would have more evaporation and the latent heat would provide more energy for disturbances. This claim is based on a number of obvious mistakes (though the claim continues to be repeated by those who should know better). *In point of fact, for Sir John's claim to even be partially true, that warmer world would have to be accompanied by lower relative humidity thus reducing the positive feedback that makes possible the prediction of warming in excess of about $1°C$ for a doubling of CO_2.*'

27. Slide 13 of 'CLIMATE CHANGE *some aspects of consensus*' by Richard S. Lindzen states: 'My last example is one where there is, indeed, consensus, but where the consensus is barely mentioned. *Kyoto will have no discernible impact on global warming regardless of what one believes about climate change.*'

28. Slide 14 of 'CLIMATE CHANGE *some aspects of consensus*' by Richard S. Lindzen states: 'A number of participants have invested so much of their reputations on an alleged concern over climate change, that it is unlikely that they will be moved by such facts. However, I think they would do well to keep the following in mind when contemplating their future image. *George Orwell wrote that language becomes "ugly and inaccurate because our thoughts are foolish, but the*

slovenliness of language makes it easier for us to have foolish thoughts." There can be little doubt that the language used to convey alarm has been sloppy at best. *It was Joseph Goebbels who said that if you repeat a lie often enough, people will believe it.* Forgetting outright lies for the moment, there is little question that repetition makes people believe things for which there is no basis.'

29. About half of current carbon dioxide emissions are being absorbed by the ocean and by land ecosystems. The threat that the terrestrial biosphere could stop being a sink for emissions and become a source of carbon was first raised by Peter Cox and co-authors in a paper in *Nature* in 2002 (Cox, P.M. et al., 'Acceleration of global warming due to carbon-cycle feedbacks in a coupled climate model', 2000: *Nature*, Vol. 408, pp. 184–87). Further research, presented by Geoff Jenkins in Moscow, showed how these carbon-cycle feedbacks had the potential to accelerate climate change from a combination of changes in vegetation carbon and soil carbon. This was published as Jones, C.D., P.M. Cox, R.L.H. Essery, D.L. Roberts, M.J. Woodage, 2003: 'Strong carbon cycle feedbacks in a climate model with interactive CO2 and sulphate aerosols', *Geophysical Research Letters*, Vol. 30, doi:10.1029/2003GL016867.

30. From the abstract of Kininmonth's paper provided in the brown paper parcel: 'The purpose of this article is to demonstrate that natural processes have caused the climate change and it is unlikely that human influences will dominate the natural processes. Any suggestion that implementation of the Kyoto Protocol will avoid future infrastructure damage, environmental degradation and loss of life from weather and climate extremes is a grand delusion.' This was published

as Kininmonth, W., 2003: 'Climate change: a natural hazard', *Energy and Environment*, Vol. 14, pp. 215–32.

31. https://en.wikipedia.org/wiki/2004_ Wimbledon_Championships_-_ Women%27s_Singles

32. See https://www.americanscientist. org/article/the-shrinking-glaciers- of-kilimanjaro-can-global-warming- be-blamed ; https://www.huffpost. com/entry/kilimanjaro-climate- change_b_1612864?guccounter=1 ; https://www.seattletimes.com/seattle- news/kilimanjaro-not-a-victim-of- climate-change-uw-scientist-says/

33. From *Flooding concept called off – New facts from the Maldives*, copy of presentation of Nils-Axel Mörner at Russian Academy of Sciences as provided in the brown paper parcel.

34. https://en.wikipedia.org/wiki/ Lysenkoism

35. This charge was repeated by Andrei Illarionov at the subsequent press conference with reference to Reiter where Illarionov was quoted as saying, 'The next point brings us directly to the Kyoto Protocol, or more specifically, to the ideological and philosophical basis on which it is built. That ideological basis can be juxtaposed and compared, as Professor Reiter has done just now, with man-hating totalitarian ideology with which he had the bad fortune to deal during the 20th century, such as National Socialism, Marxism, Eugenics, Lysenkovism [sic] and so on. All methods of distorting information existing in the world have been committed to prove the alleged validity of these theories. Misinformation, falsification, fabrication, mythology, propaganda. Because what is offered cannot be qualified in any other way than myth, nonsense and absurdity.' As reported in Remarks by Presidential Economic Adviser Andrei Illarionov at a Press Conference on Results of the Climate Change and Kyoto Protocol

Seminar in Moscow, The Federal News Service – Official Kremlin International News Broadcast, 9 July 2004. Also reported at https://www.sysecol2.ethz.ch/pdfs/Illarionov_Interv._9.Jul.04.pdf

36. https://www.sysecol2.ethz.ch/pdfs/Illarionov_Interv._9.Jul.04.pdf

37. As reported by the *Times Higher Education Supplement* on 4 June 2014: 'Armed tax police have raided British Council Offices in Moscow and eight of its other centres across Russia in a dispute over earnings from English-language teaching. The move, which is an apparent breach of the council's diplomatic protection, has prompted fears of a major diplomatic row and put a question mark over the future of one of the British Council's three flagship programmes.' See https://www.timeshighereducation.com/news/british-council-offices-raided/189281.article

38. Edition of the *Daily Mail*, Friday 9 July 2004. The article concluded: 'The link between the burning of fossil fuels and global warming is a myth. It is time the world's leaders, their scientific advisers and many environmental pressure groups woke up to the fact.'

5 In harm's way

1. A description of this heatwave that occurred during holiday season when many physicians and family members were away is provided at https://en.wikipedia.org/wiki/2003_European_heat_wave

2. The death toll of over 70,000 was calculated by Robine, J-M., S.L.K. Cheung, S. Le Roy, H. Van Oyen, C. Griffiths, J-P. Michel, F.R. Herrmann, 2008: 'Death toll exceeding 70,000 in Europe during the summer of 2003', *Comptes Rendus Biologies*, Vol. 331, pp. 171–78.

3. See https://www.geo.uio.no/edc/droughtdb/edr/DroughtEvents/_2003_Event.php for an analysis of the 2003 drought including impacts on mountainous areas and https://www.swissinfo.ch/eng/thirsty-cows_drought-affected-alpine-pastures-get-emergency-help/44305780 for information on impacts of drought on Swiss Alpine dairy herds.

4. Myles Allen is Allen, M.R., 2003: 'Liability for climate change', *Nature*, Vol. 421, pp. 891–92.

5. Luterbacher, J., D. Dietrich, E. Xoplaki, M. Grosjean, H. Wanner, 2004: 'European seasonal and annual temperature variability, trends, and extremes since 1500', *Science*, Vol. 303, pp. 1499–1503.

6. The Working Group I Contribution to the IPCC's Fourth Assessment Report, which I was invited to join as lead author by co-chair Susan Solomon, is available here: https://www.ipcc.ch/report/ar4/wg1/. An overview of the IPCC is provided at https://www.ipcc.ch/about/

7. https://www.ictp.it

8. Our finished chapter is available at https://www.ipcc.ch/site/assets/uploads/2018/02/ar4-wg1-chapter9-1.pdf

9. The IPCC review process is described at https://www.ipcc.ch/site/assets/uploads/2018/02/FS_review_process.pdf

10. Quote from *Nature* press release, personal archive.

11. Quote from *Nature* press release, personal archive.

12. These were among the impacts of climate change published in the Working Group II Contribution to the IPCC's Fourth Assessment Report;

see the Summary for Policymakers:
https://www.ipcc.ch/site/assets/
uploads/2018/02/ar4-wg2-spm-1.pdf
13. *The Times*, 2 December 2004.
14. *The Independent*, 2 December 2004.

15. *Daily Mirror*, 2 December 2004.
16. *Le Monde*, 2 December 2004.
17. *Der Spiegel*, 2 December 2004.
18. *The Washington Post*, 2 December 2004.

6 Very likely due

1. https://www.theguardian.com/world/2004/oct/23/society.russia
2. A brief summary of the meeting in Paris is provided here: http://enb.iisd.org/vol12/enb12319e.html
3. The final approved version of the text that was considered during this session is the text in the section Understanding and Attributing Climate Change from page 10 to page 12 of the Summary for Policymakers at https://www.ipcc.ch/site/assets/uploads/2018/02/ar4-wg1-spm-1.pdf. The sentences were not considered in the order they appear in the final document; the statements whose approved versions appear in the shaded box near the top of page 10 were considered last, the headline statement of which was the first in that shaded box.
4. The research I was discussing was published as Allen, M.R., P.A. Stott, J.F.B. Mitchell, R. Schnur, T.L. Delworth, 2000: 'Quantifying the uncertainty in forecasts of anthropogenic climate change', *Nature*, Vol. 407, pp. 617–20, and Stott, P.A., J.A. Kettleborough, 2002: 'Origins and estimates of uncertainty in predictions of twenty-first century temperature rise', *Nature*, Vol. 416, pp. 723–26. In these papers, we showed that by detecting the fingerprints of past temperature changes we could identify whether climate models would warm up too much, or too little, in future as greenhouse gas concentrations increased. We could do this by comparing the warming a model produced in the past in response to greenhouse gas emissions with the component of past observed warming that was attributable to greenhouse gas increases. A climate model that under- or overcooked past greenhouse warming would also probably under- or overcook future greenhouse warming. This meant we had a way to calibrate climate predictions based on the observations by scaling them up or down depending on whether the model warmed up too little or too much in the past. Since then, the predictions we made at the end of the twentieth century have been shown to accurately forecast twenty-first-century warming (Allen, M.R., J.F.B. Mitchell, P.A. Stott, 2013: 'Test of a decadal climate forecast', *Nature Geoscience*, Vol. 6, pp. 243–44).
5. The potential for quantifying past warming attributable to different anthropogenic and natural factors was investigated by Stott, P.A., J.F.B. Mitchell, M.R. Allen, T.L. Delworth, J.M. Gregory, G.A. Meehl, B.D. Santer, 2006: 'Observational Constraints on Past Attributable Warming and Predictions of Future Warming', *Journal of Climate*, Vol. 19, pp. 3055–69.
6. As shown in Stott, P.A., J.F.B. Mitchell, M.R. Allen, T.L. Delworth, J.M. Gregory, G.A. Meehl, B.D. Santer, 2006: 'Observational Constraints on Past Attributable Warming and Predictions of Future Warming', *Journal of Climate*, Vol. 19, pp. 3055–69.

7. See the discussion on page 688 of the Working Group I Contribution to the IPCC's Fourth Assessment Report: https://www.ipcc.ch/site/assets/uploads/2018/02/ar4-wg1-chapter9-1.pdf

8. The original statistical formulation for attributing past temperature changes was published in 1999 as Allen, M.R., S.F.B. Tett, 1999: 'Checking for model consistency in optimal fingerprinting', *Climate Dynamics*, Vol. 15, pp. 419–34. This was subsequently developed further to take account of the fact that climate modelling centres could only make small numbers of simulations of past climate change. This limitation in computer power meant that it wasn't possible to eliminate the effects of random chaos (internal climate variability) from modelled estimates of temperature changes due to greenhouse gas increases and other factors. The new version of the attribution method was published in 2003 as Allen, M.R., P.A. Stott, 2003: 'Estimating signal amplitudes in optimal fingerprinting, part I: theory', *Climate Dynamics*, Vol. 21, pp. 477–91. It was this new variant of the method that Myles introduced at the American Geophysical Union in 1996 that was the one most widely applied at the time of the Paris meeting.

9. This was published as Huntingford, C., P.A. Stott, M.R. Allen, F.H. Lambert, 2006: 'Incorporating model uncertainty into attribution of observed temperature change', *Geophysical Research Letters*, Vol. 33, doi:10.1029/GL024831.

10. See https://archive.ipcc.ch/pdf/assessment-report/ar4/wg1/drafts/fd/ar4_fgd_wg1_spm.pdf

11. A discussion of this issue is contained in Chapter 5 of the Working Group I Contribution to the IPCC's Fourth Assessment Report: https://www.ipcc.ch/site/assets/uploads/2018/02/ar4-wg1-chapter5-1.pdf. Figure 5.1 shows time series of global annual heat content of the upper ocean (top 700 metres) in which rapid warming is seen in the 1970s and a sharp cooling is seen from 1980 to 1985. The discussion on page 392 illustrates that at the time of the Fourth Assessment Report it could not be determined definitively whether this apparent variability was real or was a spurious artefact in the data. But the suspicion had already been raised that the ocean heat content variability was spurious in a paper by a Hadley Centre team of which I was part and which was published in 2004 (Gregory, J.M., H.T. Banks, P.A. Stott, J.A. Lowe, M.D. Palmer, 2004: 'Simulated and observed decadal variability in ocean heat content', *Geophysical Review Letters*, Vol. 31, doi:10.1029/2004GL020258). A key paper identifying the role of XBTs in distorting ocean temperature estimates (Gouretski, V., K.P. Koltermann, 2007: 'How much is the ocean really warming?', *Geophysical Research Letters*, Vol. 34, doi:10.1029/2006GL027834) was submitted in 2006 and published earlier the same month as the Paris plenary meeting but came too late to be cited in the report. The full explanation for the spurious nature of the ocean heat content variability was determined by the time of the Fifth Assessment Report and is provided in Chapter 3 (p. 261): https://www.ipcc.ch/site/assets/uploads/2018/02/WG1AR5_Chapter03_FINAL.pdf

12. See http://news.bbc.co.uk/1/hi/sci/tech/6324425.stm

13. The *Guardian*, 3 February 2007.

14. *International Herald Tribune*, 3 February 2007.

15. *Financial Times*, 3 February 2007.

16. David Miliband quoted in the *Guardian*, 3 February 2007: https://www.theguardian.com/

environment/2007/feb/03/
frontpagenews.greenpolitics

17. As quoted at https://uk.reuters.com/
article/us-globalwarming-film-
idUKL0241863020070202

7 Court of opinion

1. The court case described in this chapter is Case CO/3615/2007 in the High Court of Justice Queen's Bench Division Administrative Court between The Queen on the application of Stuart Andrew Dimmock (claimant) and The Secretary of State for Education and Skills (defendant).

2. https://en.wikipedia.org/wiki/An_Inconvenient_Truth

3. https://en.wikipedia.org/wiki/The_Great_Global_Warming_Swindle

4. A transcript to the movie is available at https://www.scripts.com/script/an_inconvenient_truth_2787

5. See https://en.wikipedia.org/wiki/The_New_Party_(UK,_2003)

6. See https://www.theguardian.com/uk/2007/oct/14/schools.film

7. See https://www.theguardian.com/science/2005/jan/27/lastword.environment; https://www.theguardian.com/world/2005/jan/27/environment.science

8. *Merchants of Doubt* by Naomi Oreskes and Erik M. Conway (Bloomsbury), Chapter 2; https://en.wikipedia.org/wiki/George_C._Marshall_Institute

9. https://en.wikipedia.org/wiki/Global_Climate_Coalition; https://www.theguardian.com/science/2005/jan/27/lastword.environment; https://en.wikipedia.org/wiki/George_C._Marshall_Institute; https://en.wikipedia.org/wiki/Jim_Inhofe

10. See https://www.sourcewatch.org/index.php/Scientific_Alliance; https://www.theguardian.com/uk/2007/oct/14/schools.film. The director of the Scientific Alliance, Martin Livermore, was the main scientific adviser on *The Great Global Warming*

Swindle film; see https://en.wikipedia.org/wiki/Scientific_Alliance; https://climatedenial.org/2007/03/09/the-great-channel-four-swindle/

11. 'Children are the key to changing society's long-term attitudes to the environment. Not only are they passionate about saving the planet, but children also have a big influence over their own families' lifestyles and behaviour.' Alan Johnson in UK Government Press Release of 2 February 2007 as quoted at https://www.cbc.ca/news/entertainment/every-british-teen-to-see-gore-s-an-inconvenient-truth-1.645206.

12. https://www.legislation.gov.uk/ukpga/1996/56/section/406; https://www.legislation.gov.uk/ukpga/1996/56/section/407

13. Rule 35.3 of the Civil Procedure Rules for Experts and Assessors (February 2005) states: '(1) It is the duty of an expert to help the court on the matters within his expertise. (2) This duty overrides any obligation to the person from whom he has received his instruction...'

14. Paragraph 30 of Witness Statement of Professor Robert Merlin Carter, 22 May 2007.

15. Carter promoted one of the 'misleading arguments' repeatedly put forward by climate deniers which was rebutted in a 2007 Royal Society publication called 'Climate Controversies: a simple guide'. The argument made by Carter in his witness statement is misleading argument number three in that list: https://royalsociety.org/-/media/Royal_Society_Content/policy/publications/2007/8031.pdf. The

primer at Carbon Brief, 'Explainer: How the rise and fall of CO2 levels influenced the ice ages' by Zeke Hausfather provides a more detailed explanation: https://www.carbonbrief. org/explainer-how-the-rise-and-fall-of-co2-levels-influenced-the-ice-ages. An explanation of what causes ice ages was also provided in the IPCC's Fourth Assessment Report (pp. 449–50) in one of the 'Frequently Asked Questions': https://www.ipcc.ch/site/assets/ uploads/2018/02/ar4-wg1-chapter6-1. pdf

16. In the film, Gore illustrated the impact on sea level 'if Greenland broke up and melted, or if half of Greenland and half of West Antarctica broke up and melted'. That, as I stated in Witness Statement of Dr Peter Stott, 27 June 2007, would 'lead to approximately 7m of sea level rise. The melting of Greenland is a real possibility. The IPCC's Fourth Assessment Report concluded that the Greenland Ice Sheet is vulnerable to complete melting if temperatures that are projected to be reached under continuing emissions were sustained for millennia. The Fourth Assessment Report also showed that the long timescales for thermal expansion of ocean water to take place means that anthropogenic sea level rise would continue for centuries, even if greenhouse gas concentrations were to be stabilised. Therefore the risk of future sea level rise presents a very important concern on long timescales.' The scientific understanding at that time of Greenland melting is summarized in the Summary for Policymakers of the Fourth Assessment Report (https://www.ipcc.ch/site/ assets/uploads/2018/02/ar4-wg1-spm-1.pdf), where it states: 'the surface mass balance [of Greenland] becomes negative at a global average warming (relative to pre-industrial levels) in excess of 1.9°C to 4.6°C. If a negative

surface mass balance were sustained for millennia, that would lead to virtual elimination of the Greenland Ice Sheet and a resulting contribution to sea level rise of about 7m. The corresponding future temperatures in Greenland are comparable to those inferred for the last interglacial period 125,000 years ago, when paleoclimatic information suggests reduction of polar land ice extent and 4 to 6m of sea level rise.' In 2007, there was already concern about the vulnerability of the West Antarctic Ice Sheet (WAIS) to melting, with a consequent 3.3m of sea level rise. The concern has heightened considerably since then. As explained in a 2020 post by Prof. Christina Hulbe for the Carbon Brief website, 'The latest research says that the threshold for irreversible loss of the WAIS likely lies between 1.5 and 2°C of global average warming above pre-industrial levels. With warming already at around 1.1°C and the Paris Agreement aiming to limit warming to 1.5°C or 'well below' 2°C the margins for avoiding this threshold are fine indeed.' See https:// www.carbonbrief.org/guest-post-how-close-is-the-west-antarctic-ice-sheet-to-a-tipping-point. Carbon Brief has an article on sea level rise at https:// www.carbonbrief.org/explainer-how-climate-change-is-accelerating-sea-level-rise.

17. The Working Group I Contribution to the IPCC's Fourth Assessment Report found that: 'Anthropogenic forcings has likely contributed to recent decreases in Arctic sea ice extent' (Chapter 9); 'it is more likely than not that anthropogenic influence has contributed to increases in the frequency of the most intense tropical cyclones' (Chapter 9); 'it is likely that future tropical cyclones (typhoons and hurricanes) will become more intense' (Summary for Policymakers). Working Group II found that: 'Corals

are vulnerable to thermal stress and have low adaptive capacity. Increases in sea surface temperature of about 1–3°C are projected to result in more frequent coral bleaching events and widespread mortality, unless there is thermal adaptation or acclimatisation by corals' (Summary for Policymakers).

18. See https://www.carbonbrief.org/10-years-on-from-hurricane-katrina-what-have-we-learned and https://www.carbonbrief.org/major-tropical-cyclones-have-become-15-more-likely-over-past-40-years

19. Paragraph 31 of Witness Statement of Dr Peter Stott. In reality, there are complex issues involved in the difficulties of people living around Lake Chad; see https://www.theguardian.com/global-development/2019/oct/22/lake-chad-shrinking-story-masks-serious-failures-of-governance

20. Paragraph 34 of Witness Statement of Dr Peter Stott. Evidence of the effects of global warming on hurricanes has grown since the time of the IPCC's Fourth Assessment Report. A 2020 study showed that tropical cyclones have become more intense over the past four decades with the most significant increases being seen in Atlantic hurricanes (https://www.carbonbrief.org/major-tropical-cyclones-have-become-15-more-likely-over-past-40-years; Kossin, J.P., K.R. Knapp, T.L. Olander, C.S. Velden, 2020: 'Global increase in major tropical cyclone exceedance probability over the past four decades', PNAS, Vol. 117, pp. 11975–80.)

21. Paragraphs 13 to 87 of Second Witness Statement of Professor Robert Merlin Carter, 22 August 2007.

22. The Summary for Policymakers of the Working Group I Contribution to the IPCC's Fourth Assessment Report stated: 'Increases in the amount of precipitation are very likely in high latitudes, while decreases are likely in most subtropical land regions, continuing observed patterns in recent trends' (https://www.ipcc.ch/site/assets/uploads/2018/02/ar4-wg1-spm-1.pdf).

23. See https://www.standard.co.uk/hp/front/labour-is-brainwashing-pupils-with-al-gore-climate-change-film-says-father-in-court-6661494.html

24. As stated in paragraph 17 of the Claimant's Skeleton Argument for Permission for Judicial Review – 'Case Law: There are no decided cases on the approach the Courts should take to sections 406 and 407 of the Education Act 1996.'

25. See paragraph 2 of Judgment of Case: https://www.bailii.org/ew/cases/EWHC/Admin/2007/2288.html

26. From analysis of the History of the Legislation as provided in paragraphs 12 to 16 of the Claimant's Skeleton Argument for Permission for Judicial Review. See also Gillard, D., 2018: 'Education in England: a history', www.educationengland.org.uk/history

27. See paragraph 12 of Judgment: https://www.bailii.org/ew/cases/EWHC/Admin/2007/2288.html

28. This is Table SPM.2 at https://www.ipcc.ch/site/assets/uploads/2018/02/ar4-wg1-spm-1.pdf

29. Transcript of An Inconvenient Truth, https://www.scripts.com/script/an_inconvenient_truth_2787

30. From Second Witness Statement of Professor Robert Merlin Carter Paragraph 68 (Regarding 'Error 15').

31. The paper referred to by Al Gore in the movie is Monnett, C., J.S. Gleason, 2006: 'Observations of mortality associated with extended open-water swimming by polar bears in the Alaskan Beaufort Sea', Polar Biology, Vol. 29, pp. 681–87.

32. This threat has been known about for a long time thanks to the well-established reduction in Arctic sea

ice and the known importance of sea ice habitat for polar bears. A more recent NASA-led study has added more evidence that polar bears are spending less time on sea ice which is leading to them having fewer cubs: https://earthobservatory.nasa.gov/images/146023/polar-bears-struggle-as-sea-ice-declines. The Working Group II Contribution to the IPCC's Fifth Assessment Report published in 2014 concluded (in its Technical Summary) that: 'Polar bears have been and will be affected by loss of annual ice over continental shelves, decreased ice duration, and decreased ice thickness' (https://www.ipcc.ch/site/assets/uploads/2018/02/WGIIAR5-TS_FINAL.pdf).

33. See http://news.bbc.co.uk/1/hi/7037671.stm

34. Paragraph 22 of Judgment: https://www.bailii.org/ew/cases/EWHC/Admin/2007/2288.html

35. Paragraph 45 of Judgment: https://www.bailii.org/ew/cases/EWHC/Admin/2007/2288.html

36. Paragraphs 24 to 33 of Judgment https://www.bailii.org/ew/cases/EWHC/Admin/2007/2288.html

37. Paragraph 34 of Judgment: https://www.bailii.org/ew/cases/EWHC/Admin/2007/2288.html

38. Hurricane intensities: Evidence of the effects of global warming on hurricanes has grown since the time of the IPCC's Fourth Assessment Report. The 2019 IPCC report concluded: 'There is emerging evidence for an increase in annual global proportion of Category 4 or 5 tropical cyclones in recent decades' and 'Extreme sea levels and coastal hazards will be exacerbated by projected increases in tropical cyclone intensity and precipitation' (IPCC, 2019: Summary for Policymakers, *IPCC Special Report on the Ocean and Cryosphere in a Changing Climate* [Pörtner, H.O.,

D.C. Roberts, V. Masson-Delmotte, P. Zhai, M. Tignor, E. Poloczanska, K. Mintenbeck, A. Alegría, M. Nicolai, A. Okem, J. Petzold, B. Rama, N.M. Weyer (eds.)]. A 2020 study showed that tropical cyclones have become more intense over the past four decades with the most significant increases being seen in Atlantic hurricanes (https://www.carbonbrief.org/major-tropical-cyclones-have-become-15-more-likely-over-past-40-years; Kossin, J.P., K.R. Knapp, T.L. Olander, C.S. Velden, 2020: 'Global increase in major tropical cyclone exceedance probability over the past four decades', *PNAS*, Vol. 117, pp. 11975–80). Polar bear mortality: A more recent NASA-led study has added more evidence that polar bears are spending less time on sea ice which is leading to them having fewer cubs: https://earthobservatory.nasa.gov/images/146023/polar-bears-struggle-as-sea-ice-declines. The Working Group II Contribution to the IPCC's Fifth Assessment Report published in 2014 concluded (in its Technical Summary) that: 'Polar bears have been and will be affected by loss of annual ice over continental shelves, decreased ice duration, and decreased ice thickness' (https://www.ipcc.ch/site/assets/uploads/2018/02/WGIIAR5-TS_FINAL.pdf). Coral reef bleaching: The 2019 IPCC report concluded: 'Marine heatwaves have already resulted in large-scale coral bleaching events at increasing frequency (*very high confidence*) causing worldwide reef degradation since 1997, and recovery is slow (more than 15 years) if it occurs (*high confidence*)' and 'Warm water corals are at high risk already and are projected to transition to very high risk even if global warming is limited to 1.5°C' (IPCC, 2019: Summary for Policymakers. In: *IPCC Special Report on the Ocean and Cryosphere in a Changing Climate* [Pörtner, H.O.,

D.C. Roberts, V. Masson-Delmotte, P. Zhai, M. Tignor, E. Poloczanska, K. Mintenbeck, A. Alegría, M. Nicolai, A. Okem, J. Petzold, B. Rama, N.M. Weyer (eds.)].

39. The 'ocean conveyor' refers to the Atlantic Meridional Overturning Circulation (AMOC), the major current system that transports warm water from the subtropics towards higher latitudes. The 2019 IPCC report concludes that the 'AMOC will very likely weaken over the 21st century (high confidence)' (https://www.ipcc.ch/site/assets/uploads/sites/3/2019/11/04_SROCC_TS_FINAL.pdf). A study published in *Nature Geoscience* in February 2021 found consistent evidence of a decline in the AMOC with evidence for it currently being in the weakest state in over 1,000 years: https://www.

ucl.ac.uk/news/2021/feb/earths-gulf-stream-system-its-weakest-over-millennium (Caesar, L., G.D. McCarthy, D.J.R. Thornalley, N. Cahill, S. Rahmstorf, 2021: 'Current Atlantic Meridional Overturning Circulation weakest in last millennium', *Nature Geoscience*, doi: 10.1038/s41561-021-00699-z. A collapse of the AMOC is unlikely but cannot be ruled out and therefore presents an important risk of climate change; see https://www.carbonbrief.org/guest-post-could-the-atlantic-overturning-circulation-shut-down

40. https://www.theguardian.com/education/2007/oct/10/schools.uk; http://news.bbc.co.uk/1/hi/7037671.stm

41. https://www.nobelpeaceprize.org/Prize-winners/Winners/2007

8 Stolen emails

1. The blog at https://www.ucl.ac.uk/global-governance/news/2021/jan/short-history-international-climate-change-negotiations-rio-glasgow provides a short history of the COP meetings based on material from *Climate Change: A Very Short Introduction*, by Mark Maslin (Oxford University Press, fourth edition).

2. Gavin posted a blog on RealClimate that Friday about the cyber attack: http://www.realclimate.org/index.php/archives/2009/11/the-cru-hack/. An analysis of the attack is provided at http://www.theguardian.com/environment/2010/feb/04/climate-change-email-hacker-police-investigation and a perspective posted by Gavin Schmidt at RealClimate one year later is at http://www.realclimate.org/index.php/archives/2010/11/one-year-later/

3. As quoted in 'Climate emails, were they hacked or just sitting in cyberspace?', The *Guardian*, 4 February 2010, http://www.theguardian.com/environment/2010/feb/04/climate-change-email-hacker-police-investigation.

4. Ibid.

5. Ibid.

6. Ibid.

7. *Daily Express*, 2 December 2009.

8. *Sunday Telegraph*, 29 November 2009: http://www.telegraph.co.uk/comment/columnists/christopherbooker/6679082/Climate-change-this-is-the-worst-scientific-scandal-of-our-generation.html

9. http://news.bbc.co.uk/1/hi/programmes/question_time/8382037.stm; https://www.youtube.com/watch?v=yLtmrjh8IBo

10. The official information from the UN Framework Convention on Climate Change on the Copenhagen meeting is at https://unfccc.int/process-and-meetings/conferences/past-conferences/copenhagen-climate-change-conference-december-2009/copenhagen-climate-change-conference-december-2009

11. An explanation of how global average temperature records work is provided by me at https://www.metoffice.gov.uk/weather/climate/science/global-temperature-records. A list of Phil Jones' publications is provided at https://people.uea.ac.uk/p_jones/publications

12. HadCRUT has had several versions as it has been improved over time. The version at the time of the CRU hack was called HadCRUT3 (https://www.metoffice.gov.uk/hadobs/hadcrut3/). This was succeeded by HadCRUT4 (https://www.metoffice.gov.uk/hadobs/hadcrut4/) and HadCRUT5 was released in December 2020 (https://www.metoffice.gov.uk/hadobs/hadcrut5/).

13. https://en.wikipedia.org/wiki/Competitive_Enterprise_Institute; https://en.wikipedia.org/wiki/Nongovernmental_International_Panel_on_Climate_Change; *Merchants of Doubt* by Naomi Oreskes and Erik M. Conway (Bloomsbury), p. 247.

14. There is more information about the divergence issue at https://skepticalscience.com/Tree-ring-proxies-divergence-problem.htm.

15. The WMO Statement on the status of the global climate in 1999 is at https://library.wmo.int/doc_num.php?explnum_id=3460

16. Mann, M.E., R.S. Bradley, M.K. Hughes, 1998: 'Global-scale temperature patterns and climate forcing over the past six centuries', *Nature*, Vol. 392, pp. 779–87; Mann, M.E., R.S. Bradley, M.K. Hughes, 1999: 'Northern hemisphere temperatures during the past millennium: Inferences, uncertainties, and limitations', *Geophysical Research Letters*, Vol. 26, pp. 759–62. See Chapter 4.

17. There is a detailed analysis of the 'hide the decline' and '*Nature* trick' phrases at https://www.skepticalscience.com/Mikes-Nature-trick-hide-the-decline.htm

18. https://web.archive.org/web/20160122151302/http://blogs.telegraph.co.uk/news/jamesdelingpole/100017393/climategate-the-final-nail-in-the-coffin-of-anthropogenic-global-warming/

19. https://www.thegwpf.org/lord-lawson-calls-for-public-inquiry-into-uea-global-warming-data-manipulation/

20. https://www.theguardian.com/environment/2009/nov/23/climate-sceptics-bob-ward-nigel-lawson

21. An interview with Phil Jones given one year afterwards describes how he felt at that time: David Adam, 2010: 'Climate: The Hottest Year', *Nature*, Vol. 468, pp. 362–64, https://www.nature.com/articles/468362a

22. 2010: 'Closing the Climategate', *Nature*, Vol. 468, p. 345, https://www.nature.com/articles/468345a

23. https://www.sciencemediacentre.org/comments-on-climate-change-following-the-story-that-emails-at-ueas-climate-research-unit-have-been-hacked-into-2/

24. https://www.thetimes.co.uk/article/top-scientists-rally-to-the-defence-of-the-met-office-5hgd5n8jlfc

25. There is a detailed analysis of Trenberth's email at https://skepticalscience.com/Trenberth-email-scandal.html and https://skepticalscience.com/Kevin-Trenberth-travesty-cant-account-for-the-lack-of-warming.htm

26. http://news.bbc.co.uk/1/hi/sci/
    tech/8392611.stm

27. RealClimate had an article one year
    later summarizing what the website
    had done during that time: http://
    www.realclimate.org/index.php/
    archives/2010/11/one-year-later/

28. See paragraph 23 of Chapter 10 of *The
    Independent Climate Change Emails
    Review*, July 2010, http://www.cce-
    review.org/pdf/FINAL%20REPORT.
    pdf

29. https://www.theguardian.com/
    environment/2009/dec/03/leaked-
    email-uea-inquiry

30. https://webarchive.nationalarchives.
    gov.uk/20091209150729/http://
    www.metoffice.gov.uk/corporate/
    pressoffice/2009/pr20091205.html;
    over 1,000 stations were weather
    stations that had been designated by
    the World Meteorological Organization
    for use in climate monitoring.

31. A picture of the long queue to gain
    entry on the first day is at https://
    www.theguardian.com/environment/
    gallery/2009/dec/07/copenhagen-
    climate-conference-pictures

32. https://www.ucl.ac.uk/global-
    governance/news/2021/jan/short-
    history-international-climate-change-
    negotiations-rio-glasgow

33. Jones et al., 1997 (Jones, P.D., T.J.
    Osborn, K.R. Briffa, 1997: 'Estimating
    sampling errors in large-scale
    temperature averages', *Journal of
    Climate*, Vol. 10, pp. 2548–68) showed
    that reliable global trends might
    be obtained from fewer than 200
    well-maintained stations, the reason
    being the geographical coherence of
    temperature trends. Various studies
    have shown that estimated global and
    hemispheric trends vary little when
    based on limited subsets of stations.
    This has been shown by Parker et
    al., 2009 (Parker, D.E., P. Jones, T.C.
    Peterson, J. Kennedy, 2009: Comment
    on 'Unresolved issues with the

assessments of multi-decadal global
land surface temperature trends' by
Roger A. Pielke Sr. et al.) and this is
illustrated in Figure 2 in Annex B
of Memorandum Submitted by Met
Office to the Parliamentary House of
Commons Inquiry 'The disclosure
of climate data from the Climatic
Research Unit at the University of
East Anglia': https://publications.
parliament.uk/pa/cm200910/cmselect/
cmsctech/387b/38722.htm

34. A discussion of using neighbour
    checks in land temperature records
    is contained in the article by
    Carbon Brief to explain how global
    temperature records are constructed:
    https://www.carbonbrief.org/explainer-
    how-data-adjustments-affect-global-
    temperature-records

35. Sea surface temperature records need
    to take account of the different ways
    the measurements have been made
    over the years. Originally they were
    made by specialized buckets used by
    mariners for measuring sea surface
    temperatures which were all variations
    on a theme – a long metal cylinder
    insulated with a tube of rubber, a
    metal flap at the top and the whole
    contraption attached to a long length
    of rope. This would be lowered over
    the side of a ship and let sink into
    the sea, water would force its way
    through the flap and the flap would
    snap shut as the bucket was hauled
    back on deck. In the very earliest days,
    measurements were taken from an
    ordinary ship's bucket made of wood
    or canvas but later custom-made
    buckets were used instead. Information
    about the use of buckets for measuring
    sea surface temperatures including
    photographs of buckets is given at
    http://www.realclimate.org/index.
    php/archives/2008/06/of-buckets-
    and-blogs/. Buckets for measuring
    sea surface temperatures are now a
    thing of the past. Instead, ships now

measure the temperature of water in their engine intakes and thousands of buoys have been deployed specifically to monitor climate. In compiling global temperature records over the last century, the different methods for measuring temperatures need to be taken into account. It takes time for a bucket to be hauled up on deck, during which water evaporates and cools the remaining seawater. The more modern methods don't have this evaporative cooling and so show generally warmer temperatures. David Parker and Chris Folland (Folland, C.K., D.E. Parker, 1995: 'Correction of instrumental biases in historical sea surface temperature data', *Quarterly Journal of the Royal Meteorological Society*, Vol. 121, pp. 319–67) worked out the magnitude of this cooling effect, which was then taken into account in the latest sea surface temperature records. Combined with the land temperature data, they produced the HadCRUT record of global temperatures over both land and sea.

36. The latest data from HadCRUT are displayed at https://www.metoffice.gov.uk/hadobs/monitoring/dashboard.html

37. As reported directly to me by my friend but also reported at Wikipedia: https://en.wikipedia.org/wiki/2009_United_Nations_Climate_Change_Conference#Activism

38. https://www.ucl.ac.uk/global-governance/news/2021/jan/short-history-international-climate-change-negotiations-rio-glasgow

39. This photograph was widely reproduced and for example was on pages 6 and 7 of the *Observer* of 20 December 2019 and also in the *Daily Telegraph*: http://www.telegraph.co.uk/news/earth/copenhagen-climate-change-confe/6841838/Copenhagen-climate-summit-world-leaders-agree-deal-but-concede-it-does-not-go-far-enough.html

40. https://science.sciencemag.org/content/suppl/2010/01/27/327.5965.510.DC1

41. https://www.theguardian.com/environment/2009/dec/18/copenhagen-deal

42. https://www.ipcc.ch/site/assets/uploads/2018/02/ar4_syr_spm.pdf

43. https://www.theccc.org.uk/wp-content/uploads/2008/12/Building-a-low-carbon-economy-Committtee-on-Climate-Change-2008.pdf

44. https://www.theccc.org.uk/2009/04/22/action-must-follow-setting-of-carbon-budgets-22-april-2009/

45. The paper was published as Stott, P.A., N.P. Gillett, G.C. Hegerl, D.J. Karoly, D.A. Stone, X. Zhang, F. Zwiers, 2010: 'Detection and attribution of climate change: a regional perspective', *Wiley Interdisciplinary Reviews – Climate Change*, Vol. 1, pp. 192–211.

46. See https://www.nature.com/articles/464141a

47. A meteorological description of that cold winter is provided at https://www.metoffice.gov.uk/binaries/content/assets/metofficegovuk/pdf/weather/learn-about/uk-past-events/interesting/2010/snow-and-low-temperatures---december-2009-to-january-2010---met-office.pdf

48. Kendon, M., 'Has there been a recent increase in UK weather records?', *Weather*, Vol. 69, pp. 327–32.

49. The articles during that week in the *Guardian* were based on the book *The Climate Files: The Battle for the Truth about Global Warming* by Fred Pearce (Guardian Books) and are to be found at http://www.theguardian.com/environment/2010/feb/03/climate-scientists-freedom-information-act

50. The *Guardian*, 2 February 2010: https://www.theguardian.com/environment/2010/feb/01/dispute-weather-fraud. A response to the

allegations that Phil had used weather stations from China that had moved thereby invalidating his results is provided at https://www.nature.com/news/2010/101115/pdf/468362a.pdf where Phil points out that an earlier statement that he was considering a correction to the allegedly flawed paper was made when he was under medication and felt under pressure to publicly concede that he had made mistakes. In this article, he also explains why the Chinese station data he used do not invalidate that paper.

51. *Guardian*, 3 February 2010. Briffa's email is cited alongside other examples to support the article's thesis that to quote the article's byline 'climate researchers manipulated the system to keep critics out of respected journals': https://www.theguardian.com/environment/2010/feb/02/hacked-climate-emails-flaws-peer-review. Why in none of the cases highlighted were anyone's views 'censored' is analysed in depth by RealClimate at http://www.realclimate.org/index.php/archives/2010/02/the-guardian-disappoints/ where they also point out that many of the same points are made by David Adam in a *Guardian* article published on 23 February 2010: https://www.theguardian.com/environment/cif-green/2010/feb/23/climate-scepticism-hacked-emails. As regards the question of whether there had been abuse of peer review, the Muir Russell report found that: 'On the allegations that there was subversion of the peer review or editorial process we find no evidence to substantiate this in the three instances examined in detail. On the basis of the independent work we commissioned (see Appendix 5) on the nature of peer review, we conclude that it is not uncommon for strongly opposed and robustly expressed positions to be taken up in heavily contested areas of science. We take the view that such behaviour does not in general threaten the integrity of peer review or publication.' (As quoted from section 1.3.3 of the report.)

52. *Guardian* editorial, Saturday 6 February 2010: https://www.theguardian.com/commentisfree/2010/feb/06/climate-science-truth-and-tribalism

53. http://www.theguardian.com/environment/georgemonbiot/2010/feb/02/climate-change-hacked-emails

54. https://whistleblower.org/politicization-of-climate-science/global-warming-denial-machine/sen-inhofe-inquisition-seeking-ways-to-criminalize-and-prosecute-17-leading-climate-scientists/

55. https://www.nature.com/articles/464141a

56. As quoted in final report; see Q103 in answer to Graham Stringer MP: https://publications.parliament.uk/pa/cm200910/cmselect/cmsctech/387/387ii.pdf

57. Stott, P.A., N.P. Gillett, G.C. Hegerl, D.J. Karoly, D.A. Stone, X. Zhang, F. Zwiers, 2010: 'Detection and attribution of climate change: a regional perspective', *Wiley Interdisciplinary Reviews – Climate Change*, Vol. 1, pp. 192–211.

58. *Guardian* 5 March 2010 (title from print edition, page 9): https://www.theguardian.com/environment/2010/mar/05/met-office-analysis-climate-change

59. https://old.parliament.uk/business/committees/committees-archive/science-technology/s-t-cru-inquiry/: Paragraph 137.

60. http://www.cce-review.org/pdf/FINAL%20REPORT.pdf: Chapter 1, paragraph 13.

61. http://www.uea.ac.uk/documents/3154295/7847337/SAP.pdf/a6f591fc-fc6e-4a70-9648-8b943d84782b: Paragraph 1 of Conclusions.

62. https://www.bbc.co.uk/news/science-environment-15373071

63. https://www.theguardian.com/theobserver/2019/nov/09/climategate-10-years-on-what-lessons-have-we-learned

64. Ibid.

65. https://www.scientificamerican.com/article/who-funds-contrariness-on/

66. https://en.wikipedia.org/wiki/Cato_Institute; https://www.cato.org/multimedia/cato-video/cato-institute-center-study-science

67. https://news.mit.edu/2009/climategate-story-1215

68. https://www.nytimes.com/2009/11/21/science/earth/21climate.html

69. https://en.wikipedia.org/wiki/Phil_Jones_(climatologist)

9 Mounting devastation

1. *The Dominion Post*, Wednesday 9 January 2013. Front Page Lead Story – 'Family of 7 escapes "tornadoes of fire"'; Above the photograph was written 'I prayed like I never prayed before'. [In my personal archive]. There is a detailed interactive account of the Holmes family experiences by the *Guardian*: https://www.theguardian.com/world/interactive/2013/may/26/firestorm-bushfire-dunalley-holmes-family

2. *The Dominion Post*, 9 January 2013.

3. Allen, M.R., D.J. Frame, C. Huntingford, C.D. Jones, J.A. Lowe, M. Meinshausen, N. Meinshausen, 2009: 'Warming caused by cumulative carbon emissions towards the trillionth tonne', *Nature*, Vol. 458, pp. 1163–66. Other papers confirming this relationship and delving further into the causes and implications followed hot on the heels of Allen et al., 2009, (submitted 25 September 2008), including Gregory et al., 2009 (submitted 24 November 2008); Matthews et al., 2009 (submitted 4 December 2008); Solomon et al., 2009 (submitted 16 December 2008): Gregory, J.M., C.D. Jones, P. Cadule, P. Friedlingstein, 2009: 'Quantifying carbon cycle feedbacks', *Journal of Climate*, Vol. 22, pp. 5232–50; Matthews, H.D., N.P. Gillett, P.A. Stott, K. Zickfeld, 2009: 'The proportionality of global warming to cumulative carbon emissions', *Nature*, Vol. 459, pp. 829–33; Solomon, S., G.-K. Plattner, R. Knutti, P. Friedlingstein, 2009: 'Irreversible climate change due to carbon dioxide emissions', *PNAS*, Vol. 106, pp. 1704–09.

4. See Guidance Note for Lead Authors of the IPCC Fifth Assessment Report on the Consistent Treatment of Uncertainties: https://www.ipcc.ch/site/assets/uploads/2017/08/AR5_Uncertainty_Guidance_Note.pdf

5. This new assessment is discussed on page 261 in Chapter 3 of the published report: https://www.ipcc.ch/site/assets/uploads/2018/02/WG1AR5_Chapter03_FINAL.pdf. The new estimates are shown in Figure SPM.3 Panel c of the Summary for Policymakers of the Fifth Assessment Report: of AR5: https://www.ipcc.ch/site/assets/uploads/2018/02/WG1AR5_SPM_FINAL.pdf

6. Stott P.A., D.A. Stone, M.R. Allen, 2004: 'Human contribution to the European heatwave of 2003', *Nature*, Vol. 432, pp. 610–14. See Chapter 5.

7. The study by Rahmstorf and Coumou found that the probability of the record heat during the Russian summer of 2010 had increased by a factor of five, meaning that 80 per cent of the risk at that time was attributable to human-

induced climate change (Rahmstorf, S., D. Coumou, 2011: 'Increase of extreme events in a warming world', *Proceedings of the National Academies of Sciences*, Vol. 108, pp. 17905–09). For the impacts of the Russian heatwave, see https://en.wikipedia.org/wiki/2010_Russian_wildfires. Two studies reached similar conclusions about the 2011 Texas heatwaves, both finding a substantial contribution from human-induced climate change to the risk of temperatures exceeding such a high threshold as was observed (Rupp, D.E., P.W. Mote, N. Massey, C.J. Rye, R. Jones, M.R. Allen, 2012: 'Did human influence on climate make the 2011 Texas drought more probable?', *Bulletin of the American Meteorological Society*, Vol. 93, pp. 1052–54; Hoerling, M., et al., 2013: 'Anatomy of an extreme event', *Journal of Climate*, Vol. 26, pp. 2811–32.) The record loss of crops during the heatwave is detailed by Hoerling et al., 2013.

8. Pall, P., et al., 2011: 'Anthropogenic greenhouse gas contribution to UK autumn flood risk', *Nature*, Vol. 470, pp. 382–85. For the heavy rainfall and flooding of autumn 2000, see https://www.metoffice.gov.uk/binaries/content/assets/metofficegovuk/pdf/weather/learn-about/uk-past-events/interesting/2000/the-wet-autumn-of-2000---met-office.pdf

9. This sentence appears in the Executive Summary of Chapter 10 of the published report at page 871: https://www.ipcc.ch/site/assets/uploads/2018/02/WG1AR5_Chapter10_FINAL.pdf

10. Lenton, T.M., H. Held, E. Kriegler, J.W. Hall, W. Lucht, S. Rahmstorf, J. Schellnhuber, 2008: 'Tipping elements in the earth's climate system', *PNAS*, Vol. 105, pp. 1766–93.

11. As summarized in the Bushfire Survival Plan I picked up with the 12 January 2013 edition of the local paper, the *Mercury – The Voice of Tasmania*, which also included a supplement *Summer Hellfire: A pictorial record*, and also as summarized in the current Fire Danger information online http://www.fire.tas.gov.au/Show?pageId=colFireDanger and in the *Guardian*'s account of the fires in Dunalley: https://www.theguardian.com/world/interactive/2013/may/26/firestorm-bushfire-dunalley-holmes-family

12. See https://www.theguardian.com/environment/damian-carrington-blog/2013/jan/08/australia-bush-fires-heatwave-temperature-scale

13. The weather conditions that day are described in the 2013 Tasmanian Bushfires Inquiry, Volume One: http://www.dpac.tas.gov.au/__data/assets/pdf_file/0015/208131/1.Tasmanian_Bushfires_Inquiry_Report.pdf

14. https://www.theguardian.com/global-development/gallery/2013/may/20/worst-natural-disasters-2012-in-pictures

15. *An Appeal to Reason: A Cool Look at Global Warming* by Nigel Lawson (Overlook Duckworth); *The Rational Optimist: How Prosperity Evolves* by Matt Ridley (Harper).

16. 'Explaining Extreme Events from a Climate Perspective', *Bulletin of the American Meteorological Society*: https://www.ametsoc.org/ams/index.cfm/publications/bulletin-of-the-american-meteorological-society-bams/explaining-extreme-events-from-a-climate-perspective/

17. This interview became the subject of complaints to the BBC; see https://www.carbonbrief.org/bbc-upholds-complaint-over-today-programme-nigel-lawson-interview. A transcript of the interview is provided by the GWPF at https://www.thegwpf.com/hoskins-vs-lawson-the-climate-debate-the-bbc-wants-to-censor/

10 False balance

1. https://en.wikipedia.org/wiki/2011_United_Nations_Climate_Change_Conference

2. https://www.ucl.ac.uk/global-governance/news/2021/jan/short-history-international-climate-change-negotiations-rio-glasgow

3. There are brief published accounts of this meeting at https://enb.iisd.org/vol12/enb12581e.html and https://enb.iisd.org/climate/ipcc36/about.html

4. As published in speeches by Lena Ek at https://www.government.se/contentassets/df02c1bf39d842a8a61809ec0748ecdd/speeches-2010-2014---lena-ek

5. The IPCC and its processes are described at https://www.ipcc.ch/about/

6. Information about the report is at http://www.climatechange2013.org

7. https://en.wikipedia.org/wiki/Industrial_Revolution

8. The Carbon Brief website provides more context to the challenge of defining the pre-industrial era at https://www.carbonbrief.org/challenge-defining-pre-industrial-era

9. https://twitter.com/hausfath/status/1349793328821243905?s=20

10. The IPCC's Fifth Assessment Report didn't formally designate the 1850–1900 average as being representative of pre-industrial temperatures but the implication was that warming trends relative to the 1850–1900 average were indicative of warming trends since pre-industrial times. The IPCC's 2018 Special Report on Global Warming of 1.5°C did formally state that the 1850–1900 period was being taken to be representative of pre-industrial temperatures. See https://www.ipcc.ch/site/assets/uploads/sites/2/2019/05/SR15_FAQ_Low_Res.pdf

11. Ridley sets out his views on alarmism in his article 'What the climate wars did to science' in the section 'Cheerleaders for alarm': https://www.rationaloptimist.com/blog/what-the-climate-wars-did-to-science.aspx. An analysis of the ways in which Ridley's views are not representative of the scientific consensus has been carried out by Carbon Brief at https://www.carbonbrief.org/scientists-respond-to-matt-ridleys-climate-change-claims. Lawson's views on alarmism are set out in his report for the GWPF, 'The Trouble With Climate Change' in the section 'Alarmism and its basis': https://www.thegwpf.org/content/uploads/2014/05/Lawson-Trouble-with-climate-change.pdf. An analysis of the ways in which Lawson's views are not representative of the scientific consensus has been carried out by Carbon Brief at https://www.carbonbrief.org/factcheck-lord-lawson-inaccurate-claims-about-climate-change-bbc-radio-four

12. The UN Office for Disaster Risk Reduction found a 151 per cent rise in direct economic losses from climate-related disasters between the 1998–2017 period and the 1978–1997 period: https://public.wmo.int/en/media/news/new-report-highlights-economic-cost-of-disasters

13. See Sutton, R., 2019: 'Climate science needs to take risk assessment much more seriously', *Bulletin of the American Meteorological Society*, Vol. 100, pp. 1637–42.

14. This is the statement on page 5 of the Summary for Policymakers of the Working Group I Contribution to the IPCC's Fourth Assessment Report: https://www.ipcc.ch/site/assets/uploads/2018/02/ar4-wg1-spm-1.pdf

15. See https://www.carbonbrief.org/
analysis-how-well-have-climate-
models-projected-global-warming

16. Page 15 of the Summary for
Policymakers of the Working Group I
Contribution to the IPCC's Fifth
Assessment Report: https://www.ipcc.
ch/report/ar5/wg1/

17. Page 19 of the Summary for
Policymakers of the IPCC's Fifth
Assessment Working Group I
Contribution to the IPCC's Fifth
Assessment Report: https://www.ipcc.
ch/report/ar5/wg1/

18. See https://enb.iisd.org/vol12/
enb12581e.html

19. This is Table 10.1 of Chapter 10, a
table that runs to eight pages (pages
932–39): https://www.ipcc.ch/site/
assets/uploads/2018/02/WG1AR5_
Chapter10_FINAL.pdf

20. In the scientific literature, the
relationship between cumulative
emissions and global temperature rise
is denoted by the metric Transient
Climate Response to Cumulative
Carbon Emissions (TCRE), which
is defined as the ratio of global
temperature rise to cumulative
emissions. The scientific finding that
TCRE is relatively constant over time
and independent of the scenario of
emissions is set out at page 1108 of
the Working Group I Contribution to
the IPCC's Fifth Assessment Report
and further explanation is provided at
pages 1112–13. A more comprehensive
explanation for why TCRE is constant
(and hence why there is a linear
relationship between temperature
and cumulative emissions) was
obtained later by MacDougall and
Friedlingstein, 2015, albeit that
their analysis confirmed the earlier
hypotheses advanced in the earlier
papers assessed in the Fifth Assessment
Report; as they summarized: 'From
the analysis above, it appears that
the hypothesis for the near-constant

nature of TCRE proposed by Matthews
et al. (2009) is largely correct. In
the more historically accurate case
of exponentially increasing CO_2
emissions... the ocean takes up a
diminishing fraction of carbon.
This leads to a stronger radiative
forcing from CO_2 at given cumulative
emissions but also increases the rate
of ocean heat uptake, compensating
in such a way as to maintain a near-
constant TCRE' (MacDougall, A.H.,
P. Friedlingstein, 2015: 'The origins
and limits of the near proportionality
between climate warming and
cumulative CO_2 emissions', *Journal of
Climate*, Vol. 28, pp. 4217–30).

21. This number takes account of the
effects of other human-induced
contributors to global warming as well
as carbon dioxide, including methane
and nitrous oxide. See page 10 of the
Summary for Policymakers of the
IPCC's 2014 Synthesis Report for a
concise statement of this finding with a
link to additional context: https://www.
ipcc.ch/site/assets/uploads/2018/02/
SYR_AR5_FINAL_full.pdf

22. A transcript of *World At One*,
27 September 2013, presenter Shaun
Ley, is available at https://sites.
google.com/site/mytranscriptbox/
home/20130927_w1

23. https://www.bbc.co.uk/bbctrust/our_
work/editorial_standards/impartiality/
science_impartiality.html

11 Citizen power

1. The Citizens' Assembly of Ireland inquiry into 'How the state can make Ireland a leader in tackling climate change' whose meeting I attended is documented here: https://2016-2018. citizensassembly.ie/How-the-State-can-make-Ireland-a-leader-in-tackling-climate-change/

2. https://en.wikipedia.org/wiki/ Citizens%27_Assembly_(Ireland); https://www.oireachtas.ie/en/debates/ debate/seanad/2016-07-15/3/

3. https://en.wikipedia.org/wiki/Mary_ Laffoy

4. The inquiry is documented at https:// 2016-2018.citizensassembly.ie/en/ The-Eighth-Amendment-of-the-Constitution/

5. Curtin's paper and presentation are at https://2016-2018.citizensassembly. ie/en/Meetings/Joseph-Curtin-Paper.pdf and https://2016-2018. citizensassembly.ie/en/Meetings/ Joseph-Curtin-Presentation.pdf

6. The Paris Agreement: https://unfccc. int/files/meetings/paris_nov_2015/ application/pdf/paris_agreement_ english_.pdf

7. https://climateactiontracker.org/ countries/eu/ assesses that the EU needs to reduce by 65 per cent by 2030 while the EU Commission recommended the target be increased to at least 55 per cent.

8. https://climateactiontracker.org/ countries/eu/; https://www.theccc. org.uk/wp-content/uploads/2019/07/ CCC-2019-Progress-in-reducing-UK-emissions.pdf

9. My PowerPoint slides are at https://2016-2018.citizensassembly.ie/ en/Meetings/Peter-Stott-Presentation. pdf

10. Fischer, E.M., R. Knutti, 2015: 'Anthropogenic contribution to global occurrence of heavy-precipitation and high-temperature extremes', Nature Climate Change, Vol. 5, pp. 560–65. I wrote a commentary on this article published as Stott, P., 2015: 'Weather risks in a warming world', Nature Climate Change, Vol. 5, pp. 516–17, https://www.nature.com/articles/ nclimate2640?platform=hootsuite

11. Varghese's presentation is at https://2016-2018.citizensassembly.ie/ en/Meetings/Dr-Saji-Varghese-Paper. pdf

12. A 2020 study published in Nature estimated global cooling at the Last Glacial Maximum to be 6.1 degrees Celsius (95% confidence interval of 5.7 to 6.5 degrees) – Tierney, J.E., J. Zhu, J. King, S.B. Malevich, G.J. Hakim, C.J. Poulsen, 2020: 'Glacial cooling and climate sensitivity revisited', Nature, Vol. 584, pp. 569–73, https://www. nature.com/articles/s41586-020-2617-x

13. The news report including the video that went out on the TV news is at https://www.rte.ie/ news/2017/0930/908678-citizens-assembly/

14. https://www.bbc.co.uk/news/ business-35119173

15. https://en.wikipedia.org/wiki/Feed-in_ tariffs_in_the_United_Kingdom; https://en.wikipedia.org/wiki/Feed-in_ tariff#Ireland

16. The final report is at https://2016-2018.citizensassembly.ie/en/ How-the-State-can-make-Ireland-a-leader-in-tackling-climate-change/ Final-Report-on-how-the-State-can-make-Ireland-a-leader-in-tackling-climate-change/Climate-Change-Report-Final.pdf

17. http://www.legislation.gov.uk/ ukpga/2008/27/contents; https:// www.theccc.org.uk/wp-content/ uploads/2008/12/Building-a-low-

carbon-economy-Committtee-on-Climate-Change-2008.pdf

18. An analysis of changing public engagement on climate change which reflects the low levels of concern and their rise in 2019 (see Chapter 12) is 'Guest post: Polls reveal surge in concern in UK about climate change' by Leo Baresi: https://www.carbonbrief.org/guest-post-rolls-reveal-surge-in-concern-in-uk-about-climate-change. Carbon Brief carried out an analysis of mentions of climate change in the UK media over the past decade: https://twitter.com/CarbonBrief/status/1123571071951757312

19. https://www.bbc.co.uk/news/science-environment-40899188

20. As quoted on BBC *Today* on 10 August 2017 in the transcript provided by the Global Warming Policy Foundation: https://www.ofcom.org.uk/__data/assets/pdf_file/0023/114674/today-programme-foi-annex.pdf

21. https://en.wikipedia.org/wiki/United_States_withdrawal_from_the_Paris_Agreement

22. https://www.theguardian.com/us-news/2018/nov/26/trump-national-climate-assessment-dont-believe; The National Climate Assessment Report is at https://www.globalchange.gov/climate-change

23. https://en.wikipedia.org/wiki/Turnbull_Government#Energy_policy

24. https://www.smh.com.au/environment/climate-change/climate-will-be-all-gone-as-csiro-swings-jobs-axe-scientists-say-20160204-gml7jy.html; https://www.theguardian.com/australia-news/2016/apr/06/csiro-climate-cuts-about-cutting-public-good-research-documents-show

25. https://www.theguardian.com/environment/2014/jan/10/david-cameron-floods-global-warming-climate-scientists

26. Met Office analysis of 2018 summer heatwave: https://www.metoffice.gov.uk/about-us/press-office/news/weather-and-climate/2019/drivers-for-2018-uk-summer-heatwave

27. The impacts of extreme weather in 2018 in South Korea are analysed in the annual report in the *Bulletin of the American Meteorological Society*, 'Explaining Extreme Events from a Climate Perspective', Chapter 18: https://journals.ametsoc.org/view/journals/bams/101/1/bams-explainingextremeevents2018.1.xml; for impacts in North East Asia see: https://en.wikipedia.org/wiki/2018_Northeast_Asia_heat_wave; the extreme wildfire season in California is analysed in the annual report in the *Bulletin of the American Meteorological Society*, 'Explaining Extreme Events from a Climate Perspective', Chapter 1: https://journals.ametsoc.org/view/journals/bams/101/1/bams-explainingextremeevents2018.1.xml for impacts in Myanmar see: https://www.humanitarianresponse.info/en/disaster/fl-2018-000124-mmr; and for drought in Scandinavia and loss of crops see: https://www.government.se/articles/2018/08/the-agricultural-sector-and-the-drought/

28. https://en.wikipedia.org/wiki/Greta_Thunberg

29. https://www.southhams-today.co.uk/article.cfm?id=117997&headline=Students%20protest%20climate%20inaction%20through%20Exeter&sectionIs=news&searchyear=2019

30. The IPCC Special Report on Global Warming of 1.5°C: https://www.ipcc.ch/sr15/

31. https://www.ipcc.ch/sr15/mulitimedia/video/

32. An accessible account by Myles of this mathematics is here: https://www.carbonbrief.org/myles-allen-can-we-hold-global-temperatures-to-1-5c

33. https://www.ipcc.ch/sr15/mulitimedia/video/

12 Change is coming

1. From the culmination of Greta Thunberg's speech to the Plenary of COP24 in Katowice: https://fridaysforfuture.org/what-we-do/activist-speeches/; https://www.youtube.com/watch?v=HzeekxtyFOY&feature=youtu.be

2. Connie Hedegaard's role at the Durban COP is described here: https://www.theguardian.com/environment/2011/dec/09/durban-climate-change-connie-hedegaard

3. For an explanation of the rule book and the outcomes of the Katowice meeting, see https://www.carbonbrief.org/cop24-key-outcomes-agreed-at-the-un-climate-talks-in-katowice

4. The analysis I presented is described here: https://www.metoffice.gov.uk/about-us/press-office/news/weather-and-climate/2018/2018-uk-summer-heatwave

5. https://www.carbonbrief.org/exclusive-bbc-issues-internal-guidance-on-how-to-report-climate-change

6. The landmark BBC documentary was previewed by the Carbon Brief website: https://www.carbonbrief.org/exclusive-bbc-one-show-first-primetime-film-climate-change-since-2007. The programme is (currently) available here https://www.bbc.co.uk/programmes/m00049b1

7. https://blog.metoffice.gov.uk/2019/05/02/a-varied-april-comes-to-a-warm-conclusion/

8. https://twitter.com/CarbonBrief/status/1123571071951757312

9. https://www.theccc.org.uk/publication/net-zero-the-uks-contribution-to-stopping-global-warming/

10. https://climate.copernicus.eu/sea-ice-cover-june-2019

11. https://timesofindia.indiatimes.com/city/delhi/delhi-records-all-time-high-of-48-degrees-celsius-heat-wave-to-continue/articleshow/69727572.cms; https://earthobservatory.nasa.gov/images/145167/heatwave-in-india

12. http://www.meteo.fr/cic/meetings/2019/IMSC/

13. The record-breaking heatwave is described at https://www.bbc.co.uk/news/world-europe-48795264

14. https://www.worldweatherattribution.org

15. https://www.bbc.co.uk/news/world-europe-48795264

16. https://climate.copernicus.eu/record-breaking-temperatures-june

17. The analysis is at https://www.worldweatherattribution.org/human-contribution-to-record-breaking-june-2019-heatwave-in-france/

18. https://blogs.scientificamerican.com/eye-of-the-storm/the-top-10-weather-and-climate-stories-of-2019/

19. https://en.wikipedia.org/wiki/List_of_Atlantic_hurricane_records

20. https://en.wikipedia.org/wiki/United_States_withdrawal_from_the_Paris_Agreement

21. https://en.wikipedia.org/wiki/Jair_Bolsonaro#Environment; https://en.wikipedia.org/wiki/2019_United_Nations_Climate_Change_Conference

22. https://www.carbonbrief.org/cop25-key-outcomes-agreed-at-the-un-climate-talks-in-madrid; https://www.newscientist.com/article/2227541-cop25-climate-summit-ends-in-staggering-failure-of-leadership/

23. https://www.carbonbrief.org/cop25-key-outcomes-agreed-at-the-un-climate-talks-in-madrid

24. Ibid.

25. https://public.wmo.int/en/media/press-release/2020-was-one-of-three-warmest-years-record

26. https://keelingcurve.ucsd.edu

27. https://www.carbonbrief.org/state-of-the-climate-2020-ties-as-warmest-year-on-record

28. https://public.wmo.int/en/media/news/reported-new-record-temperature-of-38°c-north-of-arctic-circle

29. The analysis was reported at https://www.metoffice.gov.uk/about-us/press-office/news/weather-and-climate/2020/prolonged-siberian-heat-almost-impossible-without-climate-change---attribution-study

30. https://www.youtube.com/watch?v=XUxizKqwXMw&feature=youtu.be

31. https://www.nytimes.com/2020/10/27/climate/trump-election-climate-noaa.html

32. https://www.whitehouse.gov/briefing-room/speeches-remarks/2021/01/20/inaugural-address-by-president-joseph-r-biden-jr/

33. https://www.climateassembly.uk

34. https://twitter.com/LeoHickman/status/1358722724005347331?s=20

35. https://twitter.com/LeoHickman/status/1359029408737148929?s=20

36. https://www.ucl.ac.uk/global-governance/news/2021/jan/short-history-international-climate-change-negotiations-rio-glasgow

Epilogue

1. www.climatestories.org.uk

2. IPCC, 2018: Summary for Policymakers, In: Global warming of 1.5°C. An IPCC Special Report on the impacts of global warming of 1.5°C above pre-industrial levels and related global greenhouse gas emission pathways, in the context of strengthening the global response to the threat of climate change, sustainable development, and efforts to eradicate poverty: https://www.ipcc.ch/site/assets/uploads/sites/2/2019/05/SR15_SPM_version_report_LR.pdf

3. https://www.carbonbrief.org/ccc-uk-has-just-18-months-to-avoid-embarrassment-over-climate-inaction

4. https://www.theccc.org.uk/2020/12/09/building-back-better-raising-the-uks-climate-ambitions-for-2035-will-put-net-zero-within-reach-and-change-the-uk-for-the-better/

5. https://theconversation.com/for-a-carbon-neutral-uk-the-next-five-years-are-critical-heres-what-must-happen-151708; https://www.theccc.org.uk/publication/sixth-carbon-budget/

6. The members of the Climate Change Committee are listed on pages 7–8 of https://www.theccc.org.uk/wp-content/uploads/2020/12/The-Sixth-Carbon-Budget-The-UKs-path-to-Net-Zero.pdf

7. https://www.theccc.org.uk/2020/12/09/building-back-better-raising-the-uks-climate-ambitions-for-2035-will-put-net-zero-within-reach-and-change-the-uk-for-the-better/

Index